Strategic Planning: Processes, Tools and Outcomes

LANCE D. CHAMBERS
MICHAEL A.P. TAYLOR

LONDON AND NEW YORK

First published 1999 by Ashgate Publishing

Reissued 2018 by Routledge
2 Park Square, Milton Park, Abingdon, Oxon, OX14 4RN
52 Vanderbilt Avenue, New York, NY 10017

Routledge is an imprint of the Taylor & Francis Group, an informa business

A Library of Congress record exists under LC control number: 98070148

ISBN 13: 978-1-138-34585-0 (hbk)
ISBN 13: 978-1-138-34588-1 (pbk)
ISBN 13: 978-0-429-43762-5 (ebk)

Contents

List of Figures

List of Tables

Book Structure

The layout of the book is described below:

Chapter 1 Introduction
A general introduction to the book and the work included therein.

Chapter 2 How Planning is Performed
This chapter describes the process involved when decisions are made. It also outlines some of the more common problems encountered by decision -makers.

Chapter 3 Forecasting our Future
The chapter is designed to describe and define futures forecasting and the expected results and findings from such a process.

Chapter 4 Reasons for Failure in Planning
The chapter examines reasons why planning failures are so common. There are specific circumstances, constraints, and psychologies that appear to be common in situations where failure are manifest.

Chapter 5 Why we use Models
The chapter describes a range of modelling paradigms and explains their use.

Chapter 6 Genetic Algorithms
This chapter describes, briefly, the processes employed by genetic algorithms. It also describes a number of areas to which GAs have been employed. There is also a brief history of the development of GAs; this is offered because of the relative infancy of this area of research.

Chapter 7 Where we Need to go from Here
The chapter describes what further work needs to be done in the area to ensure that the knowledge required to ease strategic decision-making processes are unearthed.

1 Introduction

Any student of history, and don't forget that any event that has occurred however recent or far in the past is history, will be able to point to fantastic successes and abysmal failures in which strategic planning was the pivotal element that provided for those successes or failures. This book concerns itself with the how and why of strategic planning and illustrates the vital role it plays in our present day-to-day lives and its potential for helping in ensuring the future viability of humanity and of the cultures and societies in and alongside which we all live.

Today there is scientific speculation as to certain whale behaviours that would indicate that they plan for the future. Whales are known to explore their seasonal feeding grounds and it is speculated that the exploration is to determine the feeding grounds capacity to keep the whales fed over the season. Can we do less than the animals alongside which we live? Can we blithely carry on into the future without planning where we wish to go and how we wish to get there? Without measuring our environments potential to continue to 'feed ourselves' as a species.

For hundreds of years societies have seen hope blossom because of changes to the environments within which they existed. This blossoming has been brought about by the ending of wars, the codification of human rights, the acceptance of new Gods, changes in policies and practices and other factor changes that were believed would generate beneficial effects for individuals, groups and societies as a whole. However; far too often these hopes have been dashed by the inability of those in power, in whatever society we are considering and at whatever level of authority, to implement efficiently or foresee potential dangers of these changes. An inability to maximise the potential benefits and minimise or eliminate the disbenefits is at the core of strategic planning failure and an ability to maximize benefits and avoid or eliminate the disbenefits has been at the core of success.

Countries, corporations and other organizations are managed by those incapable of leading or managing beyond just short periods of time and within very narrow environmental conditions and structures. Countries, organizations and families have arisen that had the capacity to soar above the common-place. They are sprinkled throughout history, yet most are no longer in existance. Why? They were conquered, went bankrupt, became obsolete, lost power and influence, they failed probably for one of only two reasons. They either became too hidebound and refused to change or failed to plan well enough for change and were then overcome by the changes that occured in the environments within which they had to operate.

What has been lacking in policy making endeavours, throughout history, has been the ability to define the potential effects of any change in the environment or society of concern if any other factors in that system are altered.[1] It can and has been argued that this is a patently false assumption and that in fact the changes that will be brought about by some specified factor movement can and has been measured and taken into account before actions are undertaken to change the status quo. We will not argue the fact that trivial and often vain attempts have been made to determine potential global effects of proposed actions. We believe that far too often the analysis is limited to being performed within the mind of a single person; an individual who is trusted to be 'good at this sort of thing', the ultimate decision-maker or some other individual with the power to create change. However; we also contend that these piece meal and ad hoc attempts are very dangerous to the development of good decisions and further that truly good decision-making has been frustrated by a lack of tools to adequately assist in forecasting any such changes, by a lack of will or wisdom, by a lack of foresight, inventiveness, rational, and radical thinking and an over-supply of traditional, mechanistic and Newtonian[2] methodologies (ways of doing things).

---

[1] Who was responsible for planning the levels of assistance that the USA would give to Germany and Japan after the Second World War? Did they plan well enough? Did they realise that Germany and Japan, with the help the USA would give, would end up as rich and powerful as they are today? Did they realise that by giving the help they did that Japan would end up killing the US automotive industry, electronics industry, and beat us in significant areas of the technology race?

[2] The Newtonian model states, basically, that we can learn to understand a system if we can break the system down into its smallest constituent parts and understand those parts. If we understand the parts we understand the whole. In recent years this has proven to be a false assumption. We have at times a need to understand a system holistically, we need to understand the whole as a whole and an understanding of the parts does not help in obtaining an understanding of how the system works. For example work on creative thought in computer systems is an area where the continual breaking down of the operations of the brain has failed to

The complexity of societies, environments, systems and constructs are such that it has been impossible, before now, to determine *a priori* (before the event) all important consequences. What is required is a tool, or set of tools, that allows us to construct a model of the system under consideration, to define the inter-relationships between the factors that comprise that system and to then allow for the reporting on the effects on the system being modeled of any changes, independent of where the change is manifest, to that system.

These models do exist (the 'World Model' from the Club of Rome was one such[3], the Trends Integration Procedure (TIP) is another) although to-date they still require quantitative (mathematical) measures to drive them and are not easy or intuitive to construct. Most models employed today require massive amounts of data as inputs, require extensive amounts of up-front detailed work to obtain these data and the accuracy of such data is still questionable. However; a statement such as; 'Vehicle usage will decline if the price of fuel increases' is a universally accepted truism and is not inaccurate nor will it change over time (as long as fuel doesn't become a luxury good - from an economic perspective).[4]

The problems arise when a computer model or a person who is responsible for the construction of a model attempts to interpret a statement such as 'Vehicle usage will decline if the price of fuel increases', and to then measure the effect of a fuel price increase on vehicle usage. If we can develop systems that require information or data that can be easily described and input then we can build robust and useful models.

---

produce the breakthrough that will allow us to construct computers that can think. More and more detailed studies of the brain will not help, we are required to think holistically about thought and thought processes.

[3]We have heard it said that the Club of Rome was a failure because their forecasted futures failed to eventuate. We disagree completely. Our perspective is that the findings of the Club of Rome in fact caused individuals to totally rethink the mechanism employed to manipulate specific factors within our environment. The findings of the Club of Rome caused changes so dramatic in the world environment that the model itself could not forecast the changes that it itself brought about. The model brought about its own failure by creating a future it could not forecast. Who could have foreseen the tremendous upswell of public opinion, changes in government and corporate objectives and concerns? The Club of Rome was in all probability the seed for much of these changes in our environment. For more information on Systems Dynamics (the methodologies employed by the Club of Rome, see works by Forrester).

[4] In economic terms a luxury good increases in sales as the price goes up. For example: if the price of Rolls Royces fell then so could their level of sales. The assumption being that people are willing to pay for the prestige brought about by the high price.

From a planning perspective is it important that we know that a certain section of road will become grid-locked during the 6:45 to 7:00 am time period or is it sufficient that we know that the link will saturate during the morning peak period? Do we need to know that the construction of 32 more long-term vehicle parking bays in 'Building X' will bring about traffic conflicts in the access street or is it sufficient to know that we are very close to some catastrophic cusp? Is it important for a business to know it will sell exactly Y numbers of its products or that the number that will sell will be significantly above breakeven. In many instances exact numbers are not required, rather indications or trends are sufficient for useful and valid decision-making.[5]

The potential problems confronting the modelling profession have been understood for a number of years as the following quote indicates;

> disenchantment has arisen because ... the models failed to answer the questions posed. (Starkie, 1974)

the same article concludes,

> The infusion of at least some new methodological ideas might provide the necessary encouragement for the general reappraisal of existing ... models.

The inability of models to answer newly articulated policy questions is also recognized (Atkins, 1987) and given the rate of change of public expectations, technology, social understanding and a myriad of other factors it becomes increasingly obvious why our existing quantitative models fail to provide answers. According to Coates (1985) the objectives of futures modelling are:

- To assist in understanding the dynamics of a system,
- To identify driving trends,

[5] Let's take education for example. There is a general acceptance that the quality of education, especially in high schools is declining. We cannot, with ease, develop a quantitative measure of the decline, but is such a measure needed? We think not! The simple fact that quality has gone down is sufficient for us to know that we now need to increase the quality of education. By how much? Again without a measure we cannot, with ease, state 'how much', but it is sufficient to know that it needs to be improved. Knowing that fact is enough for us to start implementing actions to redress the decline. There are many other examples that could have been used to demonstrate that sometimes there is no need for highly accurate quantitative data to support decision making.

- Generate images of the future,
- Define opportunities, threats and problems,
- Define the full scope of alternative actions and
- Evaluate potential consequences.

these were expanded upon by Tydeman (1987):

- Prediction and forecasting
- Exploration and describing
- Planning and anticipation and
- Engineering change or prescribing best or better futures.

Given the rate of change of systems it becomes almost impossible to collect data, develop, calibrate and execute a strategic model before the model itself becomes redundant because changes to the real-world system have gone beyond the models capacity to evaluate the new and changed factors and their potential impacts.

The models available to us are complex beasts, not easily modified to incorporate new data, knowledge and understanding. Not simply changed to answer new and previously unthought of questions. Flexible models need to be developed that are simple to create and modify; so that the shortcomings of the traditional models can be overcome.

A strategic model requires a number of capabilities:

- Sufficient flexibility to answer any question (or set of questions) that relate to the system being modeled,
- Quick and easy specification, development and modification,
- Subjective/intuitive/conceptual and traditional quantitative input capabilities,
- The ability to quickly test a wide range of what-if questions,
- Forecasting capabilities,
- Results that allow for the development of actionable decisions and
- The ability to compute a hierarchy of factor sensitivities.

Questions such as "What would the effect on violent crime rates be if we increased the use of automatic security systems in financial institutions?" need to be answered. But along with the answer to this particular question the model also needs to supply information on any peripheral effects brought about by this increase in security such as "What will happen to the cost of home insurance rates if there is an increase in use of automatic security in financial institutions?". The changes in police morale, demand for prison space, demand for Central City policing, changes in greenhouse gas emissions, international trade, standards of living, health, education, etc. In other words, the model must develop a global picture of the consequences of any proposed action or change in the environment.

Many may say that it is impossible to determine changes in levels of international trade if there is a change in the levels of security in financial institutions. We would argue differently. If it is possible to construct a model that has included in its structure links between these factors, then it is possible to forecast these changes. Others will say that even if such a model could be built, that the links are so tenuous as to be meaningless and therefore any effort expended to construct such a model would be a waste of resources. We contend that until the model is constructed and the idea tested how can anyone make such a statement?

A complex system is like a bowl of jelly. If you push at one place the whole lump of jello will move. Each part is connected to every other part. It isn't possible to find any part of the jello that can be pushed, pulled, cut or prodded that will not cause the whole lump of jello to shake and move. In this context then ANY change in our environment creates change in every other part.

What should be in question is the degree to which a change in any factor creates change in any other. It is this knowledge that will assist us in asking meaningful questions of such a model, it is this knowledge that will help us isolate the factors that create the greatest harm and those that create the greatest good. It will help us make the right decisions for the future.

For too long have models disregarded the rest of the environment in which they operate. This lack is understandable. Given the degree of detail at which models generally operate it would be a laughable decision to build a model that determined the impact on the countries economy, the effect on tourism or a wealth of other similar factors if we built a new by-pass road or introduced a higher entrance requirements for a university course in agricultural science.

However, policy formulation is an amorphous undertaking that doesn't benefit from highly detailed models of the type described above. We often do not have clear cut answers to clear cut questions/problems. We need to develop systems that allow for the input of broad concepts, ideas and beliefs about the environment of concern. Output from such a model defines 'pictures' of the solution or of a forecasted environment, given these ideas and beliefs.

If we consider the environment within which we operate as a globe (figure 1.1) and that various sub-environments are described by the balloons surrounding the central globe. In transport, for example, we model transport by applying change to a particular set of transport factors. With the result being as shown in figure 1.2. Our models are capable of simulating only parts of the transport sector and so the transport balloon is deformed with sections of the transport environment itself not being modeled in any way and hence being left unaffected by the model and very little, if any, of the rest of the world system being modeled. There is only in rare cases an open conduit to the outside world and this conduit is usually very narrow and pointed.

The modelling environment required is as shown in figure 1.3. Open conduits from and to all environmental factors and a model that takes cognizance of the total environment in question.

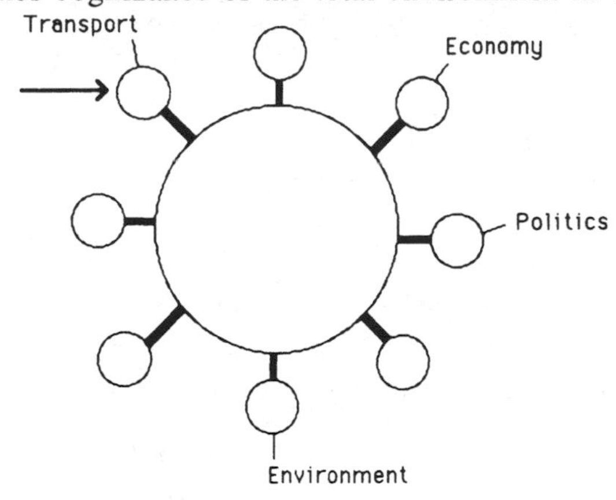

Figure 1.1 Modelling System

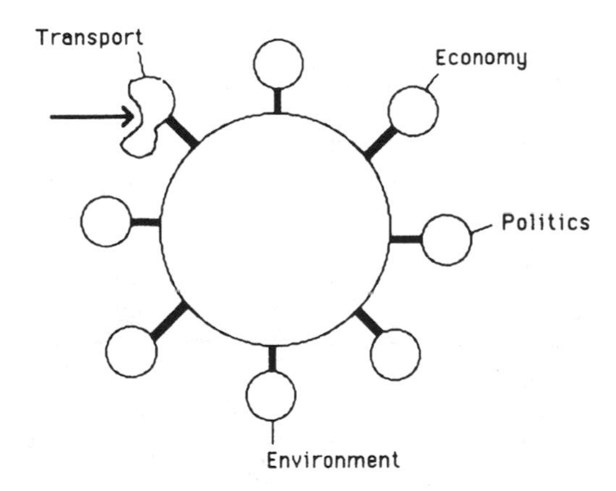

Figure 1.2 Effector on One Factor

Such a modelling system is not impossible to develop. There is a trade off between detail and depth. The greater the detail of a model then the shallower the depth, for the same level of effort, and conversely the greater the depth the lower the detail.

We propose models of great depth and low detail. Most models have concentrated, almost exclusively, on detail. This allows us to make statements such as "32 parking bays (don't forget the error term and there certainly is one) will cause traffic conflicts". They have not allowed us to ask questions of the type posed earlier. "What will be the effects on tourism if we increase the sales tax on new vehicles?"

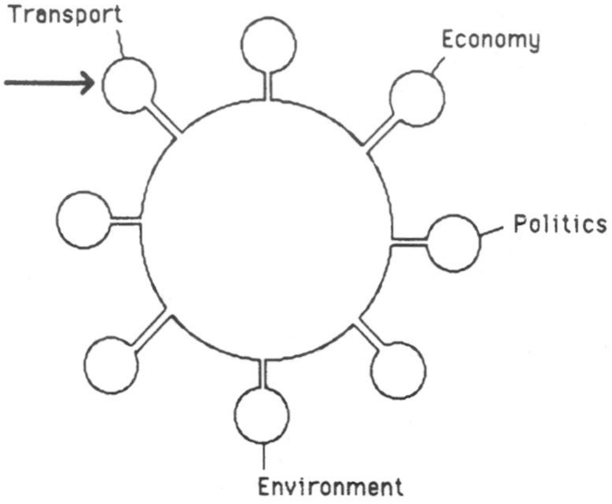

Figure 1.3 Open Modelling System

What we have not been able to do is answer off-the-cuff questions concerning proposed policy decisions or to recommend policy decisions with a comprehensive understanding of the effects of such policies on the wider ranging factors comprising the total environment upon which the model is operating and upon which the model was constructed.

It would not be sensible, nor is it suggested, that the current crop of detailed quantitative models be abandoned. These models perform a valid function in the modelling arena. What is proposed is a new crop of models to augment and extend the modelling tools generally available solidly into the strategic planning/policy formulation area.

Many policy questions are quantitative degrees of magnitude different from the models often employed to answer the questions posed.[6] It is possible to employ a detailed model of a particular

[6] We remember from the early days of our student lives the following: "You can never generate an answer more accurate than the least accurate piece of data employed." (i.e. If we have 100 data elements of 20 decimal place accuracy and 1 of 1 decimal place accuracy, then the final answer, given we use all these data, should only be given to one decimal place of accuracy). Why employ data of unit levels of

individuals spending patterns to then use this micro simulation to develop a model of the workings of the countries or maybe the worlds economy; but would one ever do it? Why then do we employ models that depend upon detailed data that is difficult and expensive to collect, collate and employ, to answer questions that are far beyond the aggregation level of the data used?

Who cares that a strategic model specifies 12,342 individuals (a number we know is wrong) will transfer from private car to public transport, at equilibrium, if the Northern Rail Spur line is built? We know that equilibrium will never occur, the environment changes too rapidly for the modeled equilibrium to come about in the real world. We need to know that there will be increases in public transport patronage and the magnitude of the increase (small, medium or large), reductions in congestion, demand for road space and airborne vehicle pollutants (and their relative magnitudes), etc. These are the questions that are needed to be answered and fictitious detailed results[7], such as 12,342 extra public transport patrons, will not answer these questions.

We require models that are tuned to answering questions on all factors introduced into the models environment. These factors require simple and easy to define parameters and cross-factor impacts. We need an easy to build world model and one that can display the system dynamics of the world we create within the model.[8]

Here is where the true value of these models comes about. Assume we are the department responsible for road funding; we can employ this type of model to forecast possible futures if we alter road funding, increase or decrease road supply, introduce traffic calming measures, increase road speeds, etc. It is possible to observe effects well outside the area of pure road works and traffic/transport impacts with these models. Changes to the factors under our immediate control allow us

---

accuracy when the answers required are in the 1 000's or 1 000 000's of unit accuracy? So we make a new statement of data accuracy needs: "Do not collect or compute data to a greater level of accuracy than the level of accuracy required for the answer."

[7]In fact such erroneous results have acted as a dis-service to modelling. When the numbers forecast do not arise people question the use of the models and unjustifiably condemn them to the scrap heap. Who should bear the brunt of the criticism should be the 'expert' model builders who cannot see what is required to answer the questions and who lack the initiative and insight required to change the tools they are constructing.

[8] There is a computer game called SimCity for the Macintosh and PC. The 'game' is being employed in universities to help teach town planning, or so we have heard and having played the game we can well believe it. The simulation is so accurate that it has been proposed that it could, in fact, be used to plan a city quite well. It is possible with very little effort to build and run a city in SimCity. Maybe, what we in the planning arena require is a sort of SimPlan!

to determine changes to factors that would normally be outside our direct sphere of influence or concern by making connections between road construction and employment, connections between vehicle usage and gasoline consumption, between vehicle pollutants and the incident's of asthma, etc.

It now becomes an imperative that these types of models be employed, not by state and federal departments, organizations or individuals in isolation, but rather by multi-disciplinary teams from as wide a range of areas as it is possible to represent. This is so that results can be viewed and a final scenario can be developed that generates global rather than local 'optima'.

These models are also valuable in helping to understand the dynamics of any given environment. Often it is impossible to predetermine the effects that a change in any one factor will have on any or all other factors. By playing what-if games with these models it becomes possible to discover the workings of the environment created within the model; that is deemed to be a true reflection of reality.

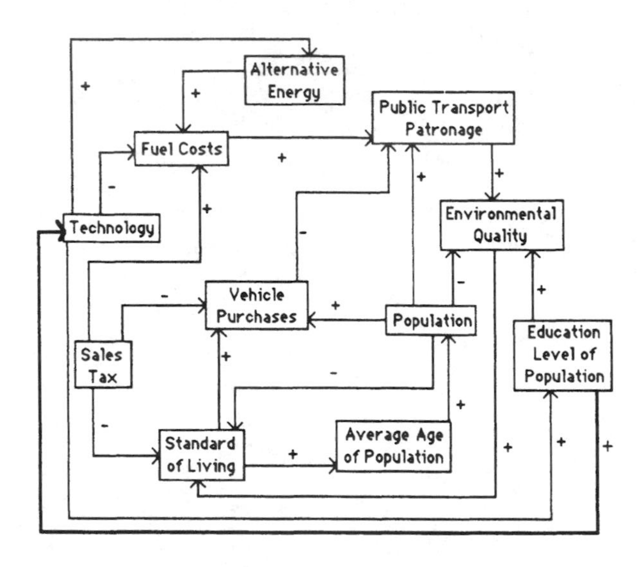

Figure 1.4 A Simple Systems Model

Figure 1.4 offers an example of a simple model that would be rather difficult to prejudge the results of given a change in any one factor, although the modelling solution system described later in the book would make short work of this modeled environment (Note that this model is not complete even within itself. The cross impact of 'Technology' vs 'Average Age of Population', the changes in population brought about by births and deaths and a number of other cross impacts are missing. A plus [+] sign on a link implies that as one factor goes up the factor it points to goes up as well and *vis-a-versa* for a minus [-] sign. For example: as the 'population' increases the 'environmental quality' goes down and as the 'standard of living' increases so will the number of 'vehicle purchases').

Table 1.1 Inter-Factor Effect Table

Factor	Direction of Effect	on Factor
Standard of Living	+	Vehicle Purchases
Vehicle Purchases	-	Public Transport Patronage
Public Transport Patronage	+	Environmental Quality
Environmental Quality	+	Standard of Living

The difficulty of determining the effects, to the whole system, of a change in any one factor can be illustrated by the use table 1.1 which is derived from the model described in figure 1.4. Let's start with 'Standard of Living'.

Let's assume the paradigm displayed above is correct and let us also assume that we experience an increase in 'Standard of Living', this causes an increase in 'Vehicle Purchases' which causes a reduction in 'Public Transport Patronage', which causes a decrease in 'Environmental Quality' which causes a decrease in 'Standard of Living' which causes a decrease in 'Vehicle Purchases' and so it goes on.

You will find, even in this simple example, many more cyclical relationships of the nature described above, some creating negative feedback, as in table 1.1, and others that create positive feedback loops that can easily spiral out of control. You will also find cycles that interact with each other and factors that are buffeted about by both positive and negative effects by various interactions in the model.

Given the real-world simplicity of the above example allied to its conceptual complexity; how can anyone be expected to understand the levels of complexity that reside in a useful and real model that could comprise thousands of factors and many more thousands of cross-impact effects with complexities that literally boggle the mind. It is obvious that no-one can understand all the ramification of a complex system and that tools must be developed to assist in this understanding.

Without this understanding it is possible to make decisions that have unexpected effects upon the environment for which decisions are being made. It is possible, as appendices 1 and 2 will show, to fail at even the primary level. That is a failure to correct the problem the decision is primarily made to eliminate. There is always the potential to also experience secondary, tertiary, etc., problems that are 'hidden' by the compexity of the situation.

'Genie', the strategic modelling tool described and offered for use later in this book can be employed in any arena that requires the application of strategic modelling techniques. We believe that this model could be employed to determine investment strategies, military policy, assist in corporate marketing decisions, national foreign policy and any other subject that can be thought of in strategic terms. These types of models can cover very small systems to total world systems (such as the Club of Rome's 'Limits to Growth' models).

It is also important that these models be able to notify the planner of factors that have large effects upon some selected factor or factors that some factor has large effect upon. In other words get the model to perform sensitivity analysis and report with a hierarchy of factors effected by each other factor and a hierarchy of factors and those they effect. With such a report available it becomes a simple task to determine the factors that need changing to bring about some required alteration in the environment and which of those factors will have the greatest desirable effect for the least effort. An example of the format of the report is given in table 1.2.

Table 1.2 Cross-Impact Effects Table

Factor	Affected By	Impact
Fuel Costs	Sales Taxes	12.04
	Interna'l Buy Price Oil	10.42
	Oil Import Levels	6.87
	Oil Reserves	-2.01

Factor	Affects	Impact
Fuel Costs	Use Public Transport	6.88
	Cost Running Vehicle	5.61
	Vehicle Purchases	-3.02

The implications of such a report are as follows:

In the first section, if we wish to increase fuel costs we will observe that the factor with the greatest impact is Sales Tax; hence, by increasing sales taxes on fuel we can have the greatest impact for the least effort. Similarly, in the second section we note that by increasing fuel costs we will increase the use of public transport to the greatest extent, followed by increasing the cost of running a vehicle and finally reducing the number of vehicles purchased.

Without a report of this form we require a system of trial-and-error to determine changes required to the system to generate a desired future. With the report we can target actions with far greater ease.

It should be noted that the report only measures primary effects. There are in these types of models secondary, tertiary, quaternary, etc., effects that are not measured or computed in this modelling paradigm, but that may have significant effects upon factors within the model. It is possible to take this general idea even further. It should be possible for a user who has generated some hypothesized future to manipulate this forecasted future into a future that is desired and to then get the model to determine which factors need modifying and to what degree, to generate this new desired future. Figure 1.5 gives an idea of what is proposed.

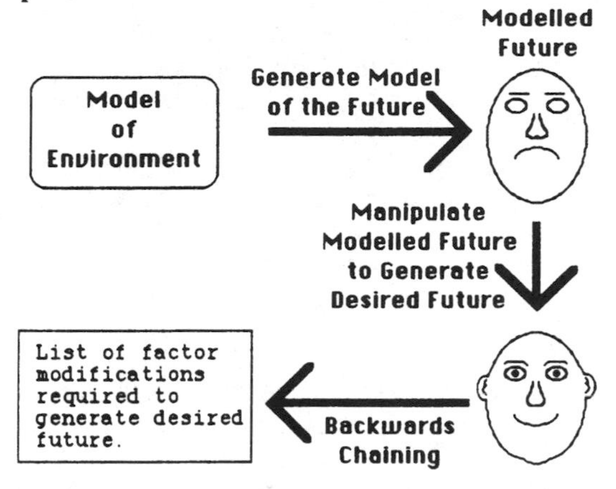

Figure 1.5 The Model Flow of Control

This modelling system now is truly a strategic policy decision-making aid. It now becomes possible for 'experts' to build the original 'Model of the Environment', to lock this model specification so that it cannot be altered and to then pass the model over to policy and decision-makers. These individuals can then manipulate the models in any way they feel necessary to achieve the required end; knowing full well the consequences of their actions. This knowledge of consequences has never truly been known in the past. It has always been possible for people to plead, and justifiably so, that the complete range of ramifications of their decisions could never be known or articulated. This is no longer true. Models of the type described will allow for total articulation and demonstration of the consequences of decisions.

One of the major difficulties in developing a model is isolating meaningful factors to be included. In any environment, there are generally a wide range of possible factors that can be employed to generate a useful model. If there is not available the resources or knowledge which of these factors are used to build the final model?

It is also possible to find a number of factor sets that will achieve the desired future. The mechanisms employed to generate a valid set of factors to be manipulated is described in greater detail in the relevant appendix (MODEL CONSTRUCTION, USE AND INTERPRETATION OF RESULTS).

Differences between Strategic and Corporate Planning

We had to do a lot of wasted reading to prepare for the writing of this book. In many cases we read books that had the term 'Strategic Planning' in their main or chapter titles but that were, as far as we are concerned, not books on Strategic but rather on Corporate Planning or Tactical Corporate Shaping.

The differences are that Strategic Planning has to do with the environment within which an organization does or will find itself, now and in the future, and with the overriding need organizations have to ensure that they remain relevant and viable in that ever changing environment. Strategic Planning has an outwards looking focus. Strategic Planners need to scan the environment looking for movements that will affect their organization and, making determinations as to what those effects will be. Determining if the effects are advantageous or detrimental, what actions can be taken to maximize benefits or minimize disbenefits of those effects. Scanning the organization to determine its potential to modify the internal or external environments so as to enhance the possibility of the organization achieving its own stated goals and objectives.

Corporate Planning has a predominantly inwards looking focus. It deals with manpower planning, budgets, training needs, etc. This is not Strategic Planning and in this book is not considered as such. A corporate planner takes the information supplied by a Strategic Planner and makes determinations on what needs to happen within the organization to achieve the visions set by the Strategic Planners.

Corporate Planners equip the organization with tactical plans to achieve the vision; Tactical Corporate Shapers ensure that the organization is physically equiped to achieve the vision but it is Strategic Planners who develop the vision to which all others in the organization work towards, because if they don't work towards that vision then the organization, will at some stage, become irrelevant in its environment and will cease to function.

We are interested in the organizations ability to fit into an environment of the future and/or measuring the organizations ability to modify that environment,[9] to enhance the organizations ability to grow, prosper and maybe change the environment for social , economic or political ends. In an attempt to put to rest the fears of 'Big Brother' being let loose with a tool of this type we would say, 'Too late!' If there is, or ever has been, a 'Big Brother' then he/she/they/it already have tools of this nature.

Where we see such a tool being used is by organizations like the Salvation Army who want to modify society by reducing the level of poverty. By constructing their 'world model' they may be able to discover more efficient ways of gaining resources and using those resources they have to more effectively bring about the changes they seek. Corporations that want to extend markets, learn how to compete, want to discover the viability of new products, governments struggling with the competing needs of environmental degradation and development, overcoming the dramatic increases in crime rates, solving problems in education, seeking new opportunities for economic cooperation, reducing tensions that can lead to war, etc.

Some of the above listed possibilities may appear far-fetched, however, we hope that as you read through this book you too will come to recognize the potential value of the system being proposed and will be able to see how it could be used to benefit you and your organization.

---

[9] Many may feel very uncomfortable with this part of the strategic planning phase. However; this is nothing new. Political parties, salesmen, corporations, journalists, religious leaders, social commentators and many other sections of society attempt to change the world – on a daily basis. Why should there be any concerns because a tool is being supplied that codifies and recognises this function implicitly.

Advertising agencies have proved to be very successful at changing the environment (society) to fulfil the goals and objectives of the corporations they represent.

2 How Planning is Performed

Some Murphisms.

Ker's Law of Planning

The most complex and inoperable plan will be implemented because it is comprehensive.

Fosters Corollary to Ker's Law

The most simple and useless plan will be implemented because it is easy to implement.

Stephen's Observations on Foster's Corollary and Ker's Laws.

The best plan will never be implemented because it is either too difficult to implement or is not comprehensive enough.

Hick's Hypothesis

Long-range plans aren't and long-range planning isn't.

What is Planning?

Urban, national, international, corporate and other systems are often very complex and probabilistic systems in which changes in activities or communications result in repercussions which can dramatically alter the system and its behaviour. We therefore confront systems that display positive feedback characteristics.[10]

This positive feedback in society can be demonstrated by the crime-poverty cycle. An individual in a poor enclave is prevented from exiting the situation in which they find themselves because of a lack of education. Because they lack skills that are required in a developed society their only alternative, to ensure survival, is to steal. Stealing becomes a profession that requires all their time, energy and effort and leaves no resources available to 'better themselves'.

[10] A simple example of positive feedback is when a microphone is held close to a speaker that is fed by the microphone. The high pitched screech is caused by sound from the speaker entering the microphone that is emitted by the speaker which enters the microphone, that is then emitted by the speaker, and so on. Around and around and with each circuit getting louder and louder.

Planning, at times, is much the same. Corporations and individuals step onto a treadmill and it becomes impossible to step off. There are no resources left, at the end of the day, for the corporation or individual to 'better themselves'. They go down fighting a pointless and useless battle. A battle that will not ensure their survival in the long term. Because they have failed to plan they cannot see the changes occurring all around them that guarantees their failure.

Although 'planning' is a generic term which is applicable to a wide variety of situations, in the context of systems, with which we are dealing here, planning is the use and development of actions and policies that will effect the systems survival in the long-term. Because planning is essentially oriented to the future, planners attempt to devise policies which can influence the development in desired directions according to the needs, expectations or wants of the community, stockholders or some other interested party as a whole (this is generally the averred rationale behind these decision-making processes. As will be seen later in the book these reasons are often subverted by personal greed, ignorance, ineptitude, ego, pride or a number of other factors).

The systems approach to problems has had a significant effect on the traditional method of planning, which was based on the narrow conception of a static overall development master plan. This view of planning has been changing, and now planning is regarded as an essentially dynamic process. A move towards the study of most environments as systems and acknowledgment of planning as a dynamic process were the underlying concepts of a new approach. Chadwick (1971) points out that:

> Planning is a conceptual general system. By creating a conceptual system independent of, but corresponding to, the real world system, we can seek to understand the phenomena of process and change, then to anticipate them, and finally to evaluate them; to concern ourselves with the optimization of a real world system by seeking optimization of the conceptual system.

Stages in Planning

The following basic stages can be identified in any planning process:

(1) Review and understanding
(2) Goal formulation
(3) Problem formulation
(4) Possible courses of action
(5) Evaluation
(6) Selection
(7) Implementation and control.

Review and Understanding

This part of a planner's activity requires an ability to understand the environment in terms of a holistic systems framework, i.e. the way in which the planner can view the make-up and operation of the system, the way in which particular sub-systems (if they exist and normally there are a large number in a real-world model) behave, react, and respond.

The effectiveness of previous planning decisions can and should be evaluated. This review of previous decisions and decision-making mechanisms has value in that it can supply knowledge of processes and procedures that do and don't work. A model, however good, is simply a guide. Any experiences gained in the environment of interest must surplant the findings of a model, unless the environment has changed significantly since that knowledge was gained and the model has incorporated those changes.

The isolation and definition of activities and factors within a system, as well as an analysis of any subsystems that are found in the environment under examination, form the primary tasks which are required at this stage in the process.

Goal Formulation

Having identified the environment and its systems and subsystems, the next task is the determination of goals that it is believed will correct the perceived problems that are manifest. However, such goals can only be set by measuring the current environments deviation from previously stated goals and objectives, as such we must first elicit and define the goals and objectives of the various players in the environment that is under review.

Because these various players often have very different and often incompatible goals, and because resources are generally limited it is considered impossible to simultaneously achieve all the goals stated by all the players. Therefore the standard approach is to develop a set of weights that are assigned to define the relative importance of each of the stated goals and objectives thus, in effect, prioritizing the goals from most to least important to achieve. Chadwick (1971) has a number of very interesting comments on this very point:

> Because the real world is constantly changing, planning must be concerned with continual change, and this means that the goals of planning will change with time and thus the policies necessary for optimization will also change; . . . the goal of

> politicians may be simply: to stay in power; that of entrepreneurs: to maximize profit.
>
> How can the socially-motivated planner reconcile his implied aims with goals such as these? ... The hard truth is that goal formulation is a difficult art, both technically, and politically: it is nonetheless essential, and the difficulties have to be faced - sometime - if a rational approach to the planning process is intended.

The traditional approach to attaining goals is for us to accept that the goals themselves are somewhat vague, of high order and are generalized statements. However, an objective, which is, in effect, an operational plan to help us in achieving a goal, must be stated more clearly and precisely.

Therefore, once the goals are defined, the objectives must then be clearly stated. This is because, it is believed that, it is impossible to determine any useful course of action if there does not exist a very clear operational plan and statement of the mechanism for achieving the objectives.

It is the achievement of the goals toward which these plans are directed, the reason for the expenditure of resources, the purposes for which the models have been developed and the end towards which everything is aimed.

Problem Formulation

The development of plans to correct the perceived environmental problems can only be started once the problems have been identified. They are identified by comparing the actual state of the environment that the model is attempting to emulate and the preferred state of that environment as expressed in the goals and objectives.

In addition to the goals and objectives formulated by governments, communities and planners there are also other bodies who would impose additions, deletions and changes and therefore require the development of new plans for particular systems or subsystems defined in the broad model.

We also confront goals and objectives at differing levels. Not only are there goals for the environment as a whole but there are also goals for subsystems within the whole. These differing levels of goals and objectives makes the process extremely complex, and time and resource consuming.

Plans

The next step is to prepare a number of solutions to the problem for evaluation. Such a selection of solutions/plans is a means of exploring different possibilities and the generation of alternatives is also a mechanism that helps planners learn about the environment being assessed and also helps in exploring processes which improve understanding of the consequences of each of the alternatives. The generation of alternative plans also requires a model which will show how the real environment would behave if the various plans were implemented.

We know that any factor movements will have repercussions on other factors that comprise the environment and upon the sub-systems within that environment; the use of models helps us to understand operational and structural relationships and to simulate the repercussions that will flow from changes we make to the environment of interest.

Thus models are used as predictive and explorative tools, to elaborate upon and to explore alternative plans of action.

Evaluation

The evaluation stage takes the set of plans that have been developed and measures the consequences of each so that they can be compared on how well each achieves the goals and objectives sought. These alternatives would be reported on by the planner, to the decision-maker/s, but only after analysis of the changes that each will create in the environment and on how the stated goals and objectives will be affected by the implementation of each plan.

The parts of any plan that are going to be used in the evaluation process depend directly on the goals and objectives identified, and although there are several evaluation techniques that have been developed, such as cost-benefit analysis, there are considerable problems of measuring costs and especially benefits.

There are also the less tangible consequences of any proposed solution and these are, at times, hard to measure, but these intangibles must be included as a part of the evaluation of any suggested alternative. How else can we isolate the 'best' alternative plan for implementation?

It is in the area of subjective evaluation that we are traditionally weakest. It is here that we need new systems for the weighting up of alternatives. Generally it is the inability to adequately value these elements that has caused the best laid plans to fail. An inability to adequately or accurately measure these factors has caused the implementation of decisions that have, in many case, acted as if these factors and their effects did not exist. "If you cannot measure a factor then it doesn't exist", was, and is, a commonly held belief.[11] We know that this is a false assumption and that, however difficult, we must discover mechanisms that allow us to, meaningfully, include such factors into our models and into our evaluation processes.

Selection

Once the various alternative plans had been tested and evaluated for their possible outcomes, the one which best achieves the stated goals is chosen to be implemented.

Implementation and Control

At this stage action would be taken to put the plan, which has been determined to be the best at the selection stage, in place and to start the process of change, if any change is required.[12]

If we recall the fact that the environment is dynamic, and that changes are constantly occurring then to monitor and maybe control these constant changes,[13] a mechanism which is also continuous must be set up to ensure that the implementation of the plan remains on course.

Any such system of control needs to make judgements on proposed changes to the plan. Because we confront a dynamic environment even the best plan will need to be changed and modified as it is in the process of being implemented. If, because of this dynamism, there are any proposed changes in the plan then these changes need to be tested by an updated model in order to measure the probable effect on the environment and see in which way the new proposals alter or improve the current plan.

---

[11] It is this belief system that permitted poisons to pollute our air, land and water. It is only today that we are now suffering the costs of actions that were taken by others who failed to measure the costs of the actions they were taking. One of the largest problems in this area is the difficulty of measuring these types of actions. What is the cost, to us all, of cutting down a single tree in the Amazon Basin? To a starving Brazilian there is no cost, only benefit. To the rest of humanity the cost may be infinite (the consequences could be the extinction of the human race), it could be $5 000 or it could be less than 1 cent. Only time will tell.

[12] The plan may be to do nothing and hence no changes need to be made.

[13] In the most sophisticated of models these changes would have been expected and incorporated into the model. Therefore the monitoring is simply to ensure that these changes are occuring as expected because if they do not occur as expected then the selected plan may not work as anticipated. Therefore, depending upon the type of model constructed they may be a need to stop certain forecasted factors from not moving as was forecasted.

In this way, planners can assess the impact of any proposals and make determinations as to how these proposed changes might better align the current implementation of the plan to the original goals selected. The same models, but updated to account for the dynamic changes in the environment, used under 'Possible courses of action', are used in this part of the total process.

As time passes, the needs and desires of the community change and the actual performance of the environment is not necessarily the one planned for it, this is because we are dealing with the occurance of probabilistic events whose behaviors cannot be defined with certainty. Consequently, the plans for the environment under study must be reviewed continuously to cope with the new situation, and so we must return to stage one and start the cycle again. This continual cyclic process is represented in Figure 2.1.

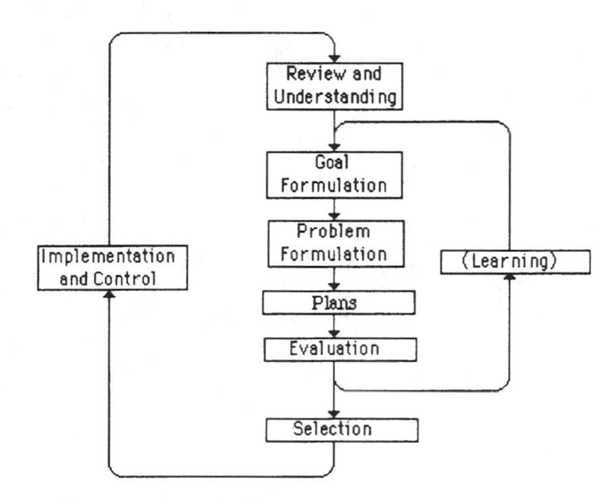

Figure 2.1 Planning as a Cyclic Process

Models in Planning

Models and modelling play an important part in planning and their use can be of significant assistance in almost all stages of the process. The understanding and analysis of environments by models, which attempt to simulate behaviors and their consequences as accurately as possible, can supply indications of where problems are arising or have potential to arise and can be of significant assistance in developing and understanding plans that can solve these problems.

Models can provide a systematic, rigorous and studied approach to understanding the relationships between different factors that comprise an environment under study. The use of models allow planners to achieve better insights into the environments being studied and hence assists in a deeper understanding of the interactive process that determine the structure of the environment that is to be altered and modified so that particular goals can be achieved.

The advantages and disadvantages of alternative plans can be assessed more easily with the help of models.

Equipped with a set of sophisticated and accurate models and evaluation techniques, a planner should be able to predict relatively accurately the consequences of various alternative actions on the environment under study. Models are the means by which planners can evaluate the probable impact that changes on factors that comprise the environment will have on that environment.

Models have been developed to describe and predict changes to almost any environment that can possibly be imagined. Some models are micro in scale, modelling only very small subsets or subsystems of much larger environments. Sometimes these models are tied together to produce macro models that are made up of a number of sub-systems which when linked together can provide a framework for the study and planning of complete cities, economies, political systems, or even for the world.

We have already pointed out that one of the greatest difficulties in the use of models lies in the evaluation of alternative plans. The measurement of benefits is still very difficult to carry out, especially if many of them are qualitative. Models are much more than straightforward forecasting methodologies; they are tools which can help planners during the implementation stage when estimating the possible effects of new developments in the plans and also with ensuring that the plans remain relevant in an ever changing environment.

The modelling techniques, predominant in the systems approach, should merely be seen as a selection of methods and techniques that can be usefully applied to any form of planning. They should also always be used with caution since in no real world situation can the model reflect the plethora of factors and effects that are possible. Therefore it is possible for a model to fail to reflect accurately the true changes that can, and often will occur in the environment for which it was constructed. Models are a very useful, but sometimes flawed, technique for helping to clarify planning problems and evaluate plans.

One final use of these models need to be recognized: the use of such models for educational purposes. Models can, and have in many cases, been developed primarily, or purely, as 'teaching machines', so that a student or planner can acquire an understanding of the environment/s they are studying.

3 Forecasting the Future

Prediction is very difficult, especially about the future.
Neils Bohr.

Introduction

When attempting to forecasting into the future, Markeley (1983) suggested some practical steps to avoid trivial forecasts, to ensure answers that require thought and interpretation and that are new, different and unique. These steps include:

• *Curiosity about change:*
Conventional theories, explanations and predictions tend to work poorly if they are not supported by a structured approach to analysis of the problem. Of central importance in a systems approach to problem resolution is a healthy curiosity about how things work, the major patterns of change that typify human history, how the human community believes it comes to understand such things, how patterns of change and of explanation may be changing and how oneself as a unique individual can learn about such things satisfactorily.

• *Gut feeling for numbers and social change:*
In addition to a knowledge of statistical analysis techniques is the need to discern patterns of meaning in quantitative and qualitative data and an ability to translate these patterns into accurate qualitative images.

• *Finding facts quickly and tooling up quickly:*
The strategic planner cannot know or have mastery over all subjects that they will need to work with. Thus an essential skill is to be able to isolate necessary facts fast, a need to be able to tolerate ambiguity and to make sense of highly unstructured data and to be able to cope with information overload. An ability to synthesise large and apparently unrelated data and information and convert it into a meaningful rule, or set of rules, that can be applied in a model.

• *Introduce complementary approaches and perspectives:*
There is no adequate theory of social change to guide the analytic methodologies of strategic planning. It is therefore necessary that planners systematically test and evaluate new approaches and perspectives.

• *An acceptance of error:*
Strategic planners must be willing and able to continually change their boundary of ideas so as to understand the changing circumstances which are manifest in the environments within which they operate. The concept of having an acceptance of error is simply to develop step to ensure that the necessary feedback from previous actions for providing valuable insights into the structure of the environment to be modeled are taken.

• *Discretion and caution regarding ideas whose time has not yet come:*
Strategic Planners, by definition, are very often working with ideas and sociological structures which are not well developed and accepted in the general society, yet it is still necessary to put the information gained from these ideas into a realistic modelling paradigm and make it available for systematic analysis.

Within a planning context for strategic forecasting, Godet (1986) outlined seven key ideas which are at the core of a stategic approach. There ideas are:

- To clarify present actions in light of the future
- To explore multiple and uncertain futures
- To adopt a global and systematic approach
- To take into account qualitative factors in the strategies of actors
- To remember always that information forecasts are not neutral
- To opt for a plurality and complementarity of approaches
- To question preconceived ideas of forecasts and forecasters.

Futuribles International, an organization founded in 1960 by Bertrand de Jouvenel is based on three central premises, premises which are believed to be at the core of strategic planning:[14]

- The future is not predetermined and is open to a plurality of possible futures which evolve over the course of time.
- The future will depend on human choices and actions, although there is no freedom of choice or actions without forecasting

[14] Definition in Futures, 18, 2, 1986, pg. 122.

activities, since there is a risk of letting things drift to the point at which we are constrained by events.

- The future is not written anywhere, rather, it is to be constructed.

From these three premises the basic role of strategic planning is to act as a 'lookout'; to monitor economic, technological, social and cultural factors, which could effect development and survival in the medium or long term. The objective is to identify the key problems that may develop in the future before they become critical and to examine what strategies can be adopted to prevent the coming about of undesirable consequences. There is also the allied task of identifying potential opportunities and developing plans to maximize benefits and achieve desired goals.

The Dangers

Today, more people live in danger of political, economic or military servitude or biological extermination than at any other time in human history. More people are exposed to common dangers, and a larger number of different perils are occurring in tandem than at any other period in our history. Threats are directed towards all forms of life and entire species are in imminent danger of extinction. Most of these hazards evolve from the ill-informed and ignorant application of technology; some from villainy, and others from ignorance, blindness, human error and lack of imagination as to potential consequences of our actions and decisions. Because technology is vital to and intimately interwoven with our cultures, institutions and social processes, intervention to deal with these dangers is the responsibility of the organizations that represent collective social choice, the governments. However, these agencies have not, as yet, shown the ability to deal with these problems in any meaningful or useful manner.

Decisions that are made by agency decision-makers become manifest in public policies which are also some of the main steering mechanisms in our society. But in the heat of actions that attempt to deal with the avalanche of day-to-day problems, sight seems to be lost of the vital role of strategic planning in bridging the chasm that exists between the present environment and the choice between a hostile future many envision or a desired future that may be preferred or proposed.

Government and Collective Security

Since one of the primary responsibilities of government is to ensure collective security, one would expect the application of effort and resources towards a continuous, concerted and informed study of the direction and manner in which we move into the future. One would expect a continual search for improved solutions, an ongoing desire for better decision-making, and efforts to create better tools and techniques to aid the processes involved.

Yet over time the hazards we confront have continued to increase in both numbers and consequence, one could almost believe that the policy-makers were deaf to any and all warnings about the future. You will find, and experience shows, that the criteria employed in determining preferred solutions to problems are heavily weighted in favor of short term achievements, without any sensible evaluation of the long-term impacts of those decisions.

The levels of will required to examine the following two crucial questions seem to be lacking in senior decision-makers and their advisors. These questions underpin policies that are concerned with the longer-term consequences of decision-making, and they are, "what will happen, if?" in those issues already being considered, and "what will happen, unless?" in those issues not being considered.

Many social analysts have stated that human society has reached such a critically balanced stage in its development that any neglect of long-term consequences to current actions or inactions will entail a penalty, for any mistakes, so expensive, distasteful, and disastrous and so harmful to the human spirit that whatever the immediate costs or inconveniences that must be borne we need to take corrective action, now, to stop current trends from presaging our destiny.

There is a need then, to change the decision-making behaviours and processes that currently exist in our institutions and are employed by our senior decision-makers. Although it would appear simple to identify the key people who must deal with these dilemmas and to exhort them to do better, the conditions of decision-making in governments and large institutions are such that circumstances may, in effect, be beyond these leaders' control. The systems that represent the world, as we know it, are becoming increasingly complex and involve many interactions among the factors that go to make up any environment of significant interest. As the systems become more complex so the ability of any one, or a group of people, to understand and handle problems in that system are eroded. It becomes impossible to make coherent and well considered decisions, the consequences of which are known of with any level of certainty.

Demands being made by constituents are, more and more, predisposed toward immediate and instant gratification of wants and these demands create increased pressures for the exercise of political expediency. A wide array of psychological as well as operational constraints may prevent leaders from dealing competently with the problems that will arise in the future. We are therefore confronted with, and are creating through our own vociferous and continuous demands, a form of social paralysis.

Whatever our beliefs in the value of strategic planning as an aid to better decision-making, we, as professionals, need to consider that much of our endeavours may simply be building a body of knowledge and an array of tools that will never be employed in any meaningful manner to solve the problems we believe it is designed to solve. If the decision-making processes themselves are disabled and incapable of operating in any useful manner then they are patently incapable of operating effectively towards the achievement of cohesive and useful goals and no selection of tools or techniques will be of value. We must first understand and then correct these dysfunctional processes.

To understand these processes requires two things. First we need to describe the present situation using tools that are superior to the subjective evidence that is available at this stage of social understanding, and second, we need to gain knowledge of the behavior patterns of institutions and decision-makers.

Decision-Makers

As we consider the situations that confront us, it is obvious that the capabilities of decision-makers are being stretched and more and more is demanded and required of them each day. The range of alternatives that are available to us and that can be implemented are greater than they have ever been. However, underlying processes and interactions of effects are more difficult to comprehend because of their sophistication and complexity. There is also the fact that there exists a new language of jargon that it must be understood before we can understand the facts themselves. If an individual cannot understand the language then it is not possible to communicate the findings or facts of the situation to them. How do you explain an 'average', 'standard deviation', 'cross-elasticity', 'carpal tunnel syndrome', etc. to an individual who has no background in the field? Sometimes there is a need to spend time 'educating' the decision-maker but they generally don't have the time. What then occurs is that the decision-maker depends upon advisors to tell them what to do and far too often these advisors are politically and ideologically biased and fail to understand, just like their masters. If the decision-makers and their advisors cannot understand the environment, problems, situation or consequences how can we expect them to implement useful and effective plans?

The largest problem we confront is that today the consequences of any error in application or omission of action can be more lethal and irreversible than they have ever been in our history. Decision-makers find themselves in a state of permanent perplexity and are at all times powerless to understand the new levels of complexity and greatly increased interdependence in our society, allied to significantly increased levels of uncertainty and risk, a continued acceleration in the pace of social, cultural and personal change, helplessness to institute the changes required to 'turn the situation around', and reductions in the relevance and efficacy of past experiences.

Experience and Loss of Value

The fact of loss of value in experience can be demonstrated by the increasing lack of respect offered the elderly, teachers, politicians and others, who were traditionally accepted as being sources of valued information in societies where these problems of discontinuity were/are not as obviously manifest.

This is sad, but true, and in part explains the loss of respect that age and experience are proffered in todays 'civilised' countries. Experience loses value in a rapidly changing environment and in fact experience may simply supply a known wrong answer to a problem. Therefore the elderly make suggestions for the solution to problems that their experience tells them are correct and the young know that the proposed answer cannot possibly be valid. The environment has changed so rapidly that answers that could solve a problem yesterday cannot solve that same problem today. However; in areas where there is little or no change, such as childbirth, you will find even the most 'modern' of women asking for their mothers advice. This is true even in cases where the mothers knowledge and experience is disregarded in almost all other issues.

The Media

Distortions in public communication, often accidental but sometimes purposeful, by the media in the never ending search for ratings and increased readerships, and also by other partisan agencies, organizations and individuals, continue to create gaps in our levels of understanding and develop friction between different parties, at many levels. These distortions inflame hate, paranoia, fear, nationalism, isolationism and tribalism and so new conflicts arise from increased levels of citizen misinformation.[15]

---

[15] Recently a news program showed a truck bursting into flame when another vehicle collided with it. The program was apparently showing up the danger faced by drivers of these vehicles in collisions. The whole story was rigged. The truck was wired to spill petroleum and a sparking mechanisms was set off to ensure the truck would burst into flame. This story later had to be retracted by the network involved, but not before it had caused a significant amount of damage to the auto

Resourcing

Resources less and less match the ever rising public expectations of the levels of service and security they expect from the agencies to which they have entrusted their futures.[16] Also, policies designed for and pushed by each, of many, narrow sectorial interest groups in the population do not operate in isolation but produce unwanted, unexpected, dangerous and erratic cross-impacts, so that in the absence of a broad perspective and cohesive thrust, there is little net gain. Each policy acts against and with the others in manners unknown and generate consequences that are unforecasted. There is no-one to take a broad prospective view and report upon the potential consequences of all these seperate, often conflicting, objectives.[17]

Decisions always, by definition, involve a choice between a number of alternatives, therefore we decide by undertaking a process of comparing the predicted consequences of each of the alternatives available and then select the best, or least-worse, alternative for implementation. That is, we determine the possible effects which we believe one decision would have now and in the future, and compare the outcome with those from the other alternatives. We endeavour to peer into the future, to make forecasts of how the environment will be changed by implementing each of the alternative actions. Our ability to understand and evaluate the possible consequences of actions is a difficult task and is highly error prone because of our general inability to peer ahead with any degree of accuracy even when the most sophisticated tools and methodologies are used, by experts and with

---

maker involved. When cases like this are made public there is really little public outcry and yet it is these very shows that have significant impacts on the shape of our societies today. There is a fear that many of our fundamental beliefs are based on lies and falsehood. William Randolph Hearst by manipulating newspapers stories, some that were outright lies, is speculated to have started the American-Mexican war (Goldberg M. H., 1990). The power inherent in the media to manipulate events is tremendous. That is why countries often enact laws that prevent excessive foreign ownership of the media. However, what do we do when our own citizens manipulate politicians and the people, for their own ends, by use of the media they own?

[16] A political commentator on the TV show '60 Minutes', Sunday ,1st Nov, 1992, mentioned that there is a continual increasing gap between what needs to be spent on social services and what the public is prepared to pay in taxes. The commentator also mentioned that if any politician was crazy enough to state that 'social services must go down or taxes must go up or a bit of both needs to happen' would probably find themselves out of office very quickly and yet one or the other must happen.

[17] The greatest saving grace is that political leaders are less and less able to initiate change, of any sort, and this includes change for the worse.

care. This is especially true in the policy and planning arenas which are the main items of discussion in this book.

Individual decision-makers, in almost all cases, want to do the right thing and to make the best decisions that they possibly can. But most of them also want to stay in office and retain their authority, to remain the CEO of the corporation, to maintain or enhance their prestige and power, to be admired by a constituency, make their lives easy, or to please the boss. It is these conflicting desires that contribute to high levels of internal conflict in many decision-makers. Because the set of objectives may not be compatible there can arise, at times, a need to make a difficult decision–to do the right thing or to survive. What can make the decision easier, at times, is the different timing that sometimes occurs between the benefits and disbenefits of such a decision. It is possible for decisions to be made, for purely personal reasons, knowing that the adverse impacts will not become obvious until many years afterwards, maybe once the individual decision-maker can no long be blamed for the failure or suffer from the consequences of that failure.[18]

The need to sometimes make trade-offs between the immediate pressures of the short term and recognition of danger of neglecting the longer term, can generate extremely high levels of stress in decision-makers and can lead to decision-making processes becoming paralyzed and dysfunctional. Often problems of this type are perceived as no-win situations and therefore avoided, abrogated, or passed onto others by the less able decision-makers. Under situations of high mental stress, we know from research of human behavior that the more immediate and short term problems always claim priority over the longer term.

The decisions, that are made in situations of high stress, are made for reasons of convenience rather than for appropriateness and decision-making behaviours become pathological in nature. In fact, the decision-making processes that are associated with making good decisions become disabled and therefore there exists no intellectual foundation upon which viable decisions can be made. Stress may be a significant factor in understanding what is happening when decision-making and decision-makers themselves become dysfunctional and often dangerous to us all.

Decision-Making and the Stress Response

When we examine policy-making processes, in any meaningful manner, it becomes necessary to closely study the behavior of

[18] Decisions about the environment fall easily and obviously into this category.

individual decision-makers. This is because there is no other way that we can achieve an understanding of the basic factors that motivate and define the general act of decision-making.

Any decision that requires a trade-off be made between a given set of choices creates anxiety in the decision-maker. Some is caused by a fear of the possiblity of adverse consequences if an error is made in the particular decision-making processes being undertaken and the level of this anxiety is directly related to the magnitude and potential damage of these consequences. Still more anxieties arise from an understanding that the reputation and self-esteem of the decision-maker are also at stake. These fears create incentives, in the decision-maker, to thoroughly search for as wide as possible selection, understanding and assessment of options and opinions in terms of these possible consequences.

Janis and Mann (date unknown) identified five patterns of decision behavior under such stress:

- Uncritical continuity of existing effort;
- Uncritical flip-flop to a completely new direction;
- Defensive avoidance or delay of a decision;
- Panic and frantic search for more and better options;
- Cool and thoughtful scanning of options and confident choice.

The first four modes are generally understood to be pathological in nature; the fifth, although never a guarantee of being error-free, is generally believed to be the most productive approach to stress.[19]

The case of uncritical continuity can be described as the classical case of bureaucratic inertia, where complacency, custom, institutional imperatives and the availability of short-term rewards for protecting past decisions dominate curiosity, vision and inventiveness. These behaviours are also characterized by institutions that have exhausted their resources, fail to learn from the past, lack incentives and lack an ability to broaden their perspective. The perceived risks brought about by change are deemed greater than the penalties of conservatism.

In the second case warnings of imminent danger are so strident and imperative that almost any movement, in any direction, or decision that attempts to change the *status quo* provides immediate relief of stress. The pressure for these impulsive types of actions are often triggered by suddenly arising events that are external to and threatening to the entire organization, or from threats that arise from within the organization that have the potential to weaken positions of power and

[19] There is however; no proof of this assertion. It is simply a generally held belief that fits in with our human perceptions of what is right. It can certainly be proved that many decisions, employing the 'pathological' processes have proved, in hindsight, to have been the correct decisions.

authority. These knee-jerk behaviours induce a short-term overreactive response to group pressures and to expediency. The decisions made do not, necessarily, make sense out of the problem or generate viable solutions. Stress levels are high for very short periods and are unsustainable.

When stress rises, as in the third category, there may be a perception of a no-win situation where the decision-maker is confronted by a 'damned if you do or damned if you don't' consequence of any decision made. Whenever feasible, in situation of this type, the decision-maker attempts to shift any responsibility away from themselves, by delegation and obfuscation, so as to minimize their personal risk. In complex organizations such shifting of responsibility is an almost routine undertaking because the persons who should have responsibility for making the decision can easily hide behind and be lost in the complexity of the organization itself.

There may also be a strong desire to delay and put off making any decision, for as long as possible, in the hope that the problem will just disappear. This tendency to delay indefinitely is a widely recognised political strategy that gains reinforcement and recognition as a valid problem solving approach because of the lack of any penalty for the manifestation of such behavior. There are also rarely penalties, at any time in the decision-makers career and only rarely during their lifetime, for ignoring future consequences of these types of behavior. Of course, if the situation is deemed hopeless and insoluble, there may be no will to respond in any way to the perceived problem.

In the fourth case, if an apparently inexorable timetable of impending penalty accompanies the perception of threat, or if new crises continue to appear on the scene, or if the resources that are capable of combating such threats begin to disappear and drain away, a sense of fear mounts. The search for answers becomes hysterical and simple and hastily contrived solutions become especially attractive; personal prejudices, biases, and beliefs tend to become more blinding with an attendant lack of critical evaluation as to the efficacy of the proposed solutions.

It is a simple task to look back over history and to isolate decisions that have been made that exhibit each of the different decision-making behaviours under stress.

Neglect of urban mass transit and intercity passenger trains falls into the first category. So did uncritical concentration of research funds on development of conventional power to the exclusion of other energy sources. The second mode is exemplified by hasty enactment of certain environmental laws, requiring for example return of rivers to a zero pollution condition, or the massive, uncritical funding of cancer research. Delays in dealing with energy policy, health care delivery, and world hunger fit the third class of decision paralysis. Over-reaction to Saddam Husseins invasion of Kuwait represents the fourth class of pathological behaviours.

With these four pathological responses to threat there is a high likelihood of error in judgement. The test of whether error actually occurs, incidentally, lies in the subsequent realisation that there was available a better alternative.[20] It is then revealed whether the decision-maker was competent to make decision in a strategic planning framework.

By comparison with the earlier four modes, the fifth is considered appealing. The threat is carefully analyzed, facts are gathered, structured and interpreted, a selection of options is examined and different consequences traced, resources are measured, and will exerted to make an 'optimal' decision.[21]

Our sense of reality tells us, however, that there are serious, comprehensive and subtle impediments to constantly achieving that desired condition. However, a commitment to rationality, at this level, is a sought after position of all policy analysts.

There are four problems that may prevent the achievement of these levels of rationality. The information gathering, understanding and dissemination processes may have any or all of the shortcomings outlined above. There are limits to human cognition and freedom from prejudice and bias which can and will impact upon understanding. Secondly, there are limits to the abilities of human imaginations to create meaningful images of the future. All kinds of social, economic, political, bureaucratic and legal pressures or constraints may be present, and rewards may be absent for the long as well as the short run. Thirdly, the intellectual and psychic resources required to make the decision may not match the situation, and physical resources to implement any such decision may be inadequate to counter the threat. Fourthly, time may be too short to think and to act.

---

[20] The ability of a decision maker is not determined by the arrival of perils or negative consequences of actions. It is determined by the ability of the decision maker to select the 'best' answer or action. It often occurs that there is no 'good' answer to a particular problem there is only the 'best' which will still create problems but minimise the consequences.

[21] Note that a decision can be to 'do nothing'. Making the correct decision does not always imply the need for change. A decision not to change is very different from the first form of pathological behaviour described, 'uncritical continuity of existing trajectories'.

Time as a Limiting Factor

The perception that time passes forwards like a treadmill, and that once an opportunity has passed we can no longer recreate that opportunity to change direction or alter a decision, may be one of the major causes of our central problem in dealing adequately with the future as is required by constructive policy development. It is believed that decision-makers are forced by a set of, external and often badly- or ill-informed, pressures and demands where the most obvious choice is to follow the line of least resistance. The decision-maker is trapped by circumstances that are dictated by the circumstances, outside their control, external to the process and highly undesirable. One of these pressures is the perception that an issue is like a time-bomb that must be immediately defused or it will explode, with devastating results. The second problem is in our perception of time as a limitation on the decision-maker who doesn't have enough of it to deal adequately with all the issues at hand. Many psychologists confirm that animals under such conditions display the symptoms of pathological behaviour that are brought on by high stress.

Clearly, there is a need to study time as it relates to decision-making processes and there are two central reasons why this is so. First, the trade off between the time available and the time required for a thourough and careful study of the situation under review, and, second, the high stress levels that can be brought about by the perceived lack of time and the added perception that this lack of time, by definition, brings with it the seeds of probable error and possible catastrophe.

In Western culture, time is generally accepted as being associated with activities that occur in a particular sequence that follow one after the other, irreversibly. It is this irreversible passing parade that creates the high levels of stress we have come to associate with decision-makers, there is significant pressure to 'get it right the first time' since we will never get another chance to visit the problem and so be able to change our decisions and actions.

There is also stress because a decision is not an action divorced from all other actions but is simply one in a series which may have been started by the perception of a threat, somewhere, a long time before. The stress is generated by the need to fit the current decision into this series, in a manner that will create benefit, but how can a decision-maker achieve this when the decision-maker is unsure if this series of past actions actually exists and if it does how does the current action fit into the series, of which they know little or nothing. There is also the possible lack of knowledge or understanding of the threat that catalyzed the series of actions in the first place or even if the threat still exists. The decision-maker is then faced with another dilema, should they attempt to understand the initial threat and then make determination upon whether actions should continue or should the system that was set in place to attack this past threat be dismantled.

The problem is obvious: if the system is dismantled and the threat still exists then the decision-maker could be making a significant mistake but if the threat is indeed past then there is the possiblility that resources, that are generally in short supply, are being wasted to no viable end. The politically easy way out is to continue as if the threat, that may be ill-understood or even totally unknown, still exists and that resources need to be brought to bear, as they have in the past, to continue the fight.[22] If the decision-maker decides to continue there will be very little backlash from any party it is only when there exists the threat that systems will be closed down or that functions stopped that the media, politicians, business people, and members of the public become concerned. So even in this regard, to avoid 'negative press', the 'best' political decision is to continue with tasks that should at best, for the rest of us, be halted and at worst scaled down.

Internal and external processes, in which the decision and its constituents are embedded, unroll at a remorseless and accelerating rate. To decision-makers time is a resource with limited availability that must be allocated in useful and usable chunks for solving of various problems. The policy-maker must budget their time for each decision, for collection and analysis of data and information, generating alternatives, evaluating the potential consequences of each alternative, and choosing from among the preferred alternatives. As will be shown, the lack of time to make rational decisions, and the unrelenting pressures of external forces, leads the decision-makers to a rapid understanding of the value of time, to the realization that time is the rarest of resources, and that a shortage in this particular resource can create conditions that lead to irrational decision-making behaviours and to the potential disasterous consequences of those decisions.

It is interesting that a public belief has grown up and even been widely promoted about how unruffled policy-makers are in their daily tasks, this fantasy may appear to be true when you see senior decision-makers on television and may also be so reported in the newspapers, but anyone who has spent time in the office of a senior official in the government is likely to report quite a different atmosphere. It can get very hectic. These people can, and often do, make compulsive

---

[22] An example of this was the discovery sometime in the 1970's of an individual that was still employed by the British government to maintain a watch on the English Channel for the arrival of Napoleons fleet and to sound the alarm if an enemy fleet was seen.
This job had continued to be filled since the beginning of the Napoleonic wars with France.

decisions to become involved in every possible activity in which they could possibly have an interest. With the interconnectedness referred to earlier, the number of issues involving any single actor is increasing. To be effective, they must take the initiative to involve themselves whenever the opportunity arises, on short notice and with timetables over which they have no control. To be sure, there are many manoeuvres available to delay or slow the action down to a more manageable pace, but generally, these people dance to tunes played and written by others. Who also dangle, metaphorically, at the end of some other string, not of their choosing.

Finally, decision-makers cannot devote their full time to contemplating matters of choice. Chief executives have numerous ceremonial functions; parlimentarians must meet constituents and help solve problems for those at home. All must offer courtesies to representatives of powerful lobby groups. All must meet the press and the public as often as possible to stay in the news. And all politicians are understandably assessing their power and prestige at every juncture, sometimes in dramatic conflict with the public interest.

Staff who support senior decision-makers play a crucial role in assisting in the decision-making process. Usually, they are under instructions to compress issue papers to one page, thus losing much of the information vital to assist in making the correct decision, because the decision-maker lacks the time to read any submission great than this in length. Given this lack of time staff at times act as surrogates for their superiors, even in cases where responsibility, much less authority, cannot be legally or morally delegated.

The hot kitchen memorialized by President Harry S. Truman exemplifies the ambient stress in the decision theater, all the more a source of torment because decision-makers, more often than not, want to do the right thing. Decision behavior under stress is thus a test not only of intelligence and judgement; it is simultaneously a test of stamina and character.

There is, however, a basic agony of choice. It is common with individuals dealing with micro-decisions and we have all, at some time, had experience with each of these dysfunctional behaviours when confronted by the situations and circumstances described. The patterns of paralysis, defensive retreat, aggression, panic or mature engagement of the problem - the five cases previously enumerated - are well known. At the macro-decision level, the same patterns apply.

Most policy decisions are close. If they were black and white, they probably would have already been made at lower levels and would not cause the levels of stress that are being described here. With decision-makers unsure because of high levels of uncertainty as to the potential consequences rather than because of a lack of responsibility, small influences, which in the normal course of events would have little impact, can nudge them either way. The last person talking to a policy-maker before a critical decision has disproportionately greater influence on the decision than any other pertinent information source. Yet as close as a choice may seem at the moment of truth, the difference in consequences years later may be enormous and even catastrophic.

The point here is that the contemplation either of uncertain but serious perils, or of long- versus short-term trade-offs, adds another creator of stress and therefore increases the chance of dysfunctional and pathological behavior.

More Points of Contention

The term strategic planning has been used as a description for a class of activities that all have one feature in common and this is the aim to discover improved methodologies for decision-making. It can be said that strategic planning is characterized by a systems approach as a preferred mechanism for obtaining a "holistic" view of the problems and environments being confronted and an understanding of problems related to decision-making. It is therefore by necessity a multi-disciplinary undertaking.

Problems in the real world give rise to the need for an integrated view of areas of knowledge which have traditionally been regarded as completely separate disciplines and have been studied and approached as such. To meet the newer requirements of strategic planning as a field of study one has to have an appreciation, for example, not only scientific, technical, social and humanistic aspects of a problem area but also, and in particular, interfaces and interconnections of such aspects with regard to changes in systems and to mechanisms which govern, motivate and direct these changes.

An understanding of the extent of particular problems and of the range of relevant factors cannot normally be gained in advance of the collection and evaluation of all relevant data and information pertinent to the problem at hand. Such an understanding grows from working on assembling and analyzing these data and information which, in the course of the process of study, are judged to be worth - or even demand - analysis.

Strategic planning is not simply characterized by efforts at prediction of the future. Forecasts of how environments will change under the effect of various forces being directed towards movements in selected factors in that environment is only one way of presenting the necessary background information required for decision-making, though admittedly such forecasts are a particularly persuasive, if not necessarily preferred, way of extracting and communicating some of the essential features of a complex condition.

There are different factors that can suggest limits in the levels of effort that it is possible and justifiable in expending when it comes to finding a basis for making a decision. Any investigation requires resources, not only economic but also a supply of qualified personnel and of much needed time. Problems of decision-making have to be solved within specific time limits that becomes an intrinsic part of the problem itself. Therefore strategic planning, more often than not, takes on the form of an exercise in the disentangling of facts, data, information, dangers, and consequences rather than the pursuit of pure research. Although it is at times difficult to maintain this distinction it is apparent that the emphasis is on trying to understand the nature and structure of a complex problem so that the decision-maker is better armed when the time comes that a decision has to be made.

There is then a difference between trying to understand the problem by using readily available information and knowledge relevant to the problem and attempts to associate the problem with areas of well-established scientific endeavour that provides a sound basis upon which to build the theoretical foundations that permit a problem to be adequately modeled.

Strategic planning, being derived from real considerations and needs, is in most cases "multidisciplinary", when seen from a purely academic scientific perspective. Study of a particular problem may therefore appear to demand a focus on areas of great urgency. In any practical situation, however, it is notoriously difficult to decide and justify in any "objective" way where to set a limit to efforts in finding, analyzing and evaluating information and knowledge that might contribute to clarify a decision-making situation. More than is readily admitted such limits are set by a vague reference to subjective factors such as experience, expectations, beliefs, attitudes, personal bias and values, sometimes joined under the collective conceptual umbrella of "judgement".

Some basic difficulties have to be considered in discussions of strategic planning as an activity and a contribution to decision-making processes - or possibly only as a support for "good judgement". These difficulties are related to the nature and validity of knowledge, and to the psychological and social factors that determine how knowledge is perceived, assimilated, evaluated and utilized in decision-making. The character of these problems can be highlighted by some observations concerning limitations in content and transfer of "information" about the future:

- Statements that can be made about future events (e.g. in analyses of alternative actions and their consequences) are in most cases affected by uncertainties whose nature cannot be clarified so as to permit conclusive assessment and judgement.

- A certain arbitrariness in delimitation of any decision-making problem is inevitable. This is valid also for the selection of information and knowledge that is selected for analysis. In addition, there are ambiguities associated with the way in which a problem is approached and interpreted, and the way in which it is thought to be related to other problems.
- Decision-makers are often dependent on advisers who carry out investigations whose results are supposed to contribute to a valuable appraisal of the decision-making situation. The adviser's experience, his intellectual capabilities, imagination, moral integrity and attitudes are difficult to assess but such an assessment is essential for an accurate assessment of the advisors results and such personal traits are an essential part of the advisors character makeup.
- Misinterpretation and misunderstanding of the real meaning and significance of presented data, information and recommendations can easily result from inadequate, inappropriate and confused communication between adviser/planner and decision-maker.
- The emotional element in decision-making is frequently such that positions taken cannot be related to a consistent system of values and norms. This problem is made worse by a lack of agreement on a set of values that is agreed to by different individuals or groups that are a party to the decision-making process.

These points categorically illustrate the necessity of being aware of the possibilities and limitations inherent in any assessment or evaluation of a particular situation. Without attempting to achieve, at least a certain minimum level of, insight and understanding which is nearly always incomplete, inaccurate, possibly biased, and difficult to verify, may even have a dysfunctional effect on decision-making processes.

Counter-productive behavior leading to self-fulfilling prophecies, and disregard of data, advice, personal observation and information because of "cognitive dissonance"[23] are but two examples of behaviours which make it difficult in some situations to understand the correct meaning and significance of any particular situation. Therefore, if the improved basis for decision-making resulting from strategic planning can result in better, more responsible decision-

[23] Cognitive dissonance is a situation where if data or information is made available to an individual and these data or information conflict with personal beliefs, perception and understanding then these data are disregarded, forgotten or held to be untrue.

making, a necessary pre-condition is a desire on the part of the decision-maker for foresight, inquisitiveness and understanding.

What is the meaning of all these terms that have been referred to? What set of procedures is it best to follow when confronted by a real life problem in order to comply with all the criteria stipulated? How is it possible to develop the level of understanding and foresight in a decision-maker which according to the arguments set forward are a prerequisite for correct action? How do values, beliefs and tacit knowledge in the decision-making process affect the validity of the final outcome? Should we not have the right to demand some measurable minimum level of moral decency and intellectual capacity in investigation, exploration, understanding and decision-making? How should we set criteria and measures in situations that required the solving of multiple goals? Can we and should we formulate "professional codes of conduct" defining, at a minimum, the responsibilities associated with accepting the positions of adviser, planner or decision-maker?

This set of questions clearly shows that the understanding and analysis of complex problems in decision-making situations can lead to problems which must be resolved if we really desire the right to claim that we know what we are talking about and that it is also of importance.

Even without a satisfactory definition of the concept of strategic planning it is easy to realize the high level of significance, of foresight, and of an understanding of alternative actions and their implications for the future. In some sense there are elements of policy formulation, planning and decision-making at every conscious level of human action. However the particulars of the factors implicit in the decision and the reasons that influence these decisions are not often thought about consciously or stated explicitly. In most situations, where a choice is not regarded as obvious and self evident, one has to consider various eventualities and think over probable or possible, desirable or undesirable consequences of each decision that is considered for implementation.

This process can lead to an, at times, tortuous self disclosure of internal conflicts because of an inherent set of personal beliefs, values and loyalties. These attributes have to be taken into account and internally resolved in one way or another before it is possible to make any decision that is in any way connected or related to these factors. This resolution does not have to be a morally acceptable one, the individual must simply come to an accommodation, whatever that accommodation may be. Increased awareness and insight results in a compelling need for scrutiny of one's beliefs, values and requires a reconsideration of goals and of means of goal-attainment.[24]

Awareness on the part of the adviser, planner and the decision-maker, of a number of factors that dominate the process of decision-making is of central importance if one is to speak about "the right decision". But such awareness is not enough.

These concerns gives rise to further questions such as: What knowledge relevant to the problem could actually be obtained and what data and information can be fed into the process? How should available resources be best employed to help elicit a "best" solution?

The decision-makers generally demonstrate by their past actions which answers they are prepared to accept in certain problem domains. These answers may differ significantly from those suggested and recommended by contemporary and later critics. But this is not, by definition, a problem with the decision-making process since there is no valid reason to look for consensus in results, since the critics are not party to the predicaments confronted by the responsible decision-maker in their role at the stated point in time. What a decision-maker should attempt to achieve is to minimize the possibility of any justified criticism being voiced because of actions they have recommended after consideration of the total situation, they must avoid it being believed that they had not fully taken into account and understood all the background information that was important to the problem at hand and that could have been made available. This statement of another of the major problems of the decision-making process may bring further understanding of the role of strategic planning in a decision-making context.

Important questions to which the parties that are involved in a decision-making process have to make up their minds, consciously or unconsciously, are treated or discussed in the following points. These questions concern strategic planning as a physical activity that planners undertake, the results of that activity and the problems of communication and responsibility. The various themes might, very briefly and without claiming to be all encompassing, be tabled as follows:

- How is it possible to make statements about the future based on knowledge and experience from the past?

---

[24] The circumstance in which a decision-making situation is perceived as trivial or a decision can be considered as self evident without the need for any preliminary analysis, certainly is not a generally valid mechanism for satisfactory or well informed and researched decision making: the impression of being "self evident" might originate from an insufficient understanding of the problem, its nature and scope, or in a lack of understanding of the levels of responsibility associated with being a good, consistent and decent decision-maker.

- What is the significance for example of a regular pattern of effects and movements such as the observance of a linear upwards trend or a cyclical change in the environment?
- With what right is it possible to claim that a model of an environment that is reasonably consistent with past experience will also give meaningful results when we use it to forecast the future?
- What limitations in the ability to describe and understand a situation or a process are inherent in the selection of an approach to solving the problem or of a particular model type?
- What salient features characterize different forecasting methods and in what situations can they be justifiably applied?
- What is the relationship between Strategic Planning and Decision Theory and how are these areas of inquiry related to Long-Range Planning?
- What are the basic methodological pitfalls associated with a long time perspective and how should they be avoided?
- How should different kinds of vaguely defined information and knowledge be assessed and assimilated in a meaningful way?
- What is the role of factors such as psychological insights, a feeling for the movement of environmental factors, and a feeling for the completeness of background information?
- How should a structured and implementable planning process be developed that allows policies to adapt to new data and information and to continually changing goals?
- How can we ensure appropriate levels of coordination when decentralized decision-making is necessary or desirable?
- What criteria are useful for evaluation of statements about the future?
- What difference does it make in the selection of evaluative criteria whether evaluation or assessment, of the effectiveness of a decision, is made before or after the event?
- What is the role of values and norms in decision-making?
- Is it possible to combine a humanist theory of value with the need for a rational approach to solutions of problems?
- What control mechanisms are needed to develop a stable and useful system of responsibilities that are shared between the decision-maker, those who delegate authority to them, and those who are affected by the decision-makers ability and capability to perform?
- What is the role, if any, of criticism in and of the decision-making processes employed in particular circumstances and how should criticism affect decision-making?

Questions like these should be included within the scope of strategic planning and should also be subjects of study and comment by strategic planners.

Underlying Causes of Shortsightedness

Using different sets of diagnostic tools and methodologies, we can conclude that in regard to long-range threats, there are numerous examples of neglect that could be discussed, and that these examples encourage a preference for the achievment of outcomes that satisfy short run needs but that also place in danger the ability of the decision system to operate effectively. If these examples of pathological behaviours, that continually favor the short run, are merely symptoms, their relief may prove superficial. Not that we expect the discovery of a simple cure to the problem. Indeed, we must be prepared to solve the complex problems by the assiduous application of complex solutions.

The painting of pictures of the multitudes of possible futures have often been reserved as the special preserve of dreamers, doomsters and science fiction writers (Ingwell, 1973). Elise Boulding (1973) describes futurists in other ways as: technocrats, social evolutionists and revolutionary futurists. Of five groups in the last category, one is defined as a "political, non-violent" approach, implicitly suggesting the general lack of political reality in other genres of peering into the future. Here, we state quite categorically that we consider strategic planning largely in terms of decision-making about the future.

But while we have focused strongly on the highest levels of policy decision-making in this work, our concern should, in reality, embrace all of the institutions that have a concern about their future survival, and on all of the forces affecting and influencing the decision-makers that make determinations on that future will be defined and constructed.

Kenneth Boulding (1964) broke significant ground here, when he stated the proposition that we stand precariously in the middle of a period of great change, where the development of a livable and enduring society will require a commitment of human energy, understanding and enlightenment far exceeding that of any of our ancestors, be they selected from any period in our history. He outlined how we, as a single cohesive race of humans, must act. We must:

- Understand how policy-makers and key institutions predict the future;
- Forecast what is possible with care and elegance;
- Imagine what possible future there might be in terms of plausible alternatives;

- Determine the opportunities to develop towards more desirable futures.

Whatever the temptation, and the temptation becomes great, to point towards some ill-defined utopia, and develop to the foreseeable, plausible and desirable, we limit this treatise, in general, to a less ambitious but perhaps from an operational and survival perspetive more critical task of dealing, primarily, with the future in terms of what we do not want, in defining futures we wish to actively and assiduously avoid. The setting of such boundaries do not limit the epistemological arena, which is very broad in nature and covers a wide array of academic fields. The capacity to imagine and foresee perils and consequences is a vital part of the needs of any society that wishes to, and has the capacity to, survive into the future, a property that C. P. Snow (1961) asserts in his statement that 'all healthy societies are ready to sacrifice the existential moment for their children's future and for children after these'.[25]

This perspective on species survival is revealed in every culture from its production of literature, art, poetry, music, science, philosophy, architecture, manufacture of consumer goods and entertainment. The perspectives inherent in the value set of individual policy-makers is inevitably molded by the social environment within which they find themselves, and it must also be remembered that the social environment that has developed was molded, in its turn, by decision-makers that came before. Understandings that arise between those that make the decisions and those who are affected by these decisions does not automatically create an atmosphere for progress, but then neither does discord and violence. The requisite atmosphere for change can be independent of these situations, they do however, effect the decision-making process but we cannot determine *a priori* what this effect will be.

With technology affecting all aspects of our society, and with the winds of conflict being blown increasingly to new areas of the globe, we must underscore one major new phenomenon, that technological decision-making is becoming more and more the prerogative of the political process. In the past the introduction of new technologies was

[25] Given the current lack of effort truly being expended on the saving of the environment, which is after all what ensures the survival of our descendents, we find Snow's contention *prima facie* evidence that our societies are no longer 'healthy'. A strong economy is but one aspect of survival yet we have reached a point where that is the single dimension at which most individuals, commercial organisations and governments wish, or even attempt, to ensure survival. As this book will hopefully demonstrate, we require more than uni-dimensional solutions to the problems we confront and the accumulation of wealth, in the form of money, is the least of these solutions.

solidly retained within the bailiwick of commerce, industry, and the military; this is no longer the case.

However, technology has tended to create environments that allowed for the concentration of power and wealth. In facilitating the explosion of the uncritical exploitation of natural resources, technology triggered an unexpected age of scarcity and produced a dramatic widening of the differences in human conditions. Hundreds of thousands starve to death while food is dumped into the oceans by others. The haves get more and the have not get less in an unending cycle of increasing polarization. The question of who wins and who loses is becoming more strident and strenuous and increasingly requires answers because of the dangers it should be communicating to those with the power and ability to change the situation.

Because the selection of ends and means is more often a matter of political choice, because publicly funded projects have become larger in scale with economically more at stake, and because government regulation limits the freedom of private enterprise, the management of technological application and introduction has inevitably become a more political process.

Other factors support this tendency towards the politicizing of technological adaptation. Technology generates more options, as with water, coal, gas, nuclear, wind, solar and geothermal sources as energy alternatives. More choices thus have to be made so attention becomes more frequently riveted on the decision event. Government tends to play a greater role in stimulation, regulation, using technology directly and investing in social overheads. So in most cases, choice is not left to the decentralized, invisible hand of the market-place; rather, social decisions are driven by circumstances, abetted by television, to high visibility and pinpointed political locale. A ratcheting then occurs in public expectations of crisis abatement, amplified by cultural trends in the abdication of individual responsibility to government. Pressures to satisfy short term expediency grow more strenuous, and, as we saw, pathological.

We repeat, decisions on technology-intensive public policy are growing more political in the sense that whatever the technical, economic, social or legal implications, more decisions are pushed upwards towards the highest level of policy authority. As these decisions continue to move up the political decision-making hierarchy they, patently, grow more political and these decisions also become more shortsighted, self serving, unsuitable and hazardous to the continued survival of the species.

Notwithstanding the disparaging inference to 'political' decisions, there is not a pragmatic basis for this situation by itself to increase risk. Political decisions are certainly not, *per se*, bad, decisions. For reasons previously advanced, they do not have to be less rational. There are two major questions that must be answered, however. Firstly, whether concentration on the short run is at the expense of the longer run in terms of consequences? And secondly, does this growing burden on the political decision-making apparatus degrade the quality of decisions made, in a general sense?

There is a view that the combination of complexity, interdependence, political workload and atmosphere of continued political conflict may render the system completely ungovernable. This perception is supported by Miles (1976) in terms of public administration. The critical test is whether the government is able to fulfil its primary responsibility: by acting as the principle guarantor of collective public security.

An alternative view of a governments role would be to consider it responsibilities in the arena of collective security against large-scale risks, and not just welfare, as the necessary, if not sufficient condition of governmental responsibility. This viewpoint answers to a reversal in the public attitude from 'let the government do it' to 'we'll do it ourselves' as far as reductions in individual risk are concerned, and an increased appreciation of, and acceptance of government intervention in, the broader and more lethal threats to survival.

Common sense, supported by a plethora of scientific and theoretical work, dictates that no political system can withstand increasing demands indefinitely, although we tend to treat the democratic political system as if it were capable of infinitely satisfying our needs and wants. History should alert us and lead us to carefully consider whether the continued ability to make the necessary tough decisions, over time, is vulnerable to instability and to situations where no meaningful or useful decisions can be made, under any circumstances.

Thus, the policy making systems are being buffeted simultaneously by more and more complex choices, by dramatic and debilitating discontinuities created by the pace of arrival of innovation and technology, by the appearance of crises and by an every increasing expansion of political responsibility. Less and less attention is being devoted to the adverse effects, on the viability of the undertaking, and to the long term impacts of these decisions on the total environment within which these decisions are made.

We are saying that the focus of attention on the short run, that is becoming more and more manifest in the decision-making process that occur, not only neglects the future itself but also potentially impacts so catastrophic that we will have no way of reversing the damage these decision may cause. It increases the possibility of future decisions being made that are wrong because the decision-making processes increasingly lag the response to threat, and are themselves seriously weakened or transported to the edge of danger of collapse under the next serious problem that arrives requiring a solution. The situation becomes ominous not only because of the potentially irreversible nature of the threats, but also, because of the magnitude of these threats and, as was said before, because neglect of the future in current decisions may undermine the long-range capability to make any sort of meaningful decisions at all

Information is the source from which political choice stems. If the system becomes hide-bound and is exhausted, early warning signals could well be masked by an intrinsic inability to observe or perceive of the danger. The slow learning curve required to be traversed to obtain the necessary intellectual ability to meet new and different situations will lag behind the arrival of dangers and pressures to act. The increasingly hectic role of policy authorities to coordinate and ease the tensions in relationships among institutionalize and parties to the crises could overshadow the primary task, which is to deal with the problems themselves.

As the stabilizing functions of government are lost, the system becomes ever more sensitive to smaller and smaller mishaps and to more trivial imbalances in the system. The sensation is like that of a novice ice-skater, obliged to devote full energies and attention simply to remaining upright, unable to muster the balance and self-confidence to engage in the creative activity of figure-skating or racing. As with humans, any defects in the hearing apparatus can cause instability and vertigo, when this happens to an institution there it is manifest when the institution itself thrashes around getting nowhere and expending tremendous energies and resource in continuing this pointless undertaking, never learning to do better, but rather getting worse as the ability to listen and understand is compromised by the continued pressures and constraints.

We now attempt to discover the underlying causes of this institutional deafness, in the policy apparatus, to warnings about the future. We look first at the malfunctioning of the decision-making process that can cause either excessive or insufficient stress for a balanced, healthy consideration of the future to take place. Candidates for such consideration are the decision-makers, our institutions and individual citizens themselves.

What we need is a change in behaviours. In our present system of minimization of coercion we tend to apply rewards and penalties where change is self-motivated rather than when it is directed from 'above'. Many observers of social behavior contend that people change far more readily and faster than do organizations, large or small. Institutional inertia in government, academia, church and industry has always been prevalent and may even be getting far more pronounced and widespread.

There is also a positive side to the levels of inertia that these institutions maintain. In every society, these institutions play a significant role in the maintenance of social norms, so that mutual roles and expectations of participants are sufficiently predictable to preserve coherence, communication conduits and understanding of particular stances. This is also, at times, construed as conservatism at its worst but whatever ones personal view on the subject of conservatism there are still a significant set of valuable precepts that fall out of this inertia. Marris (1974) has pointed out that conservatism is an impulse to defend the predictability of life 'as necessary for survival as adaptability'.

If, however, certain of our human undertakings, by nature, tend to be conservative and others innovative, mismatches and inconsistencies will arise and possibly disable the decision-making process. If, however, all the processes were conservative or all innovative we would still confront either total stagnation or total chaos. Perhaps there exists some level of difference between these two states that is essential to generating an atmosphere necessary to ensure that we are prepared for threats and are swift in finding correct and timely answer to the problems that will arise.

We do not expect that there will be any significant levels of change in our institutions and rather assume that the existing state of affairs will continue with reform, if it occurs at all, will be a very slow and tedious process.

The problem is not that decision-makers will not change it is rather that they are in situations that prevent them from doing so. Given the high priority politicians, must for reasons of political survival, place on maintaining their position, power and prestige, and given that political energies are generally targeted at the shorter run, we cannot expect a change in the processes of decision-making unless people change at a more fundamental level.

The one of the root causes of these dilemmas and a possible path to their remedy lies in developing new and useful tools that are able to forecast the future and the consequences of decisions in an easy, useful, and meaningful manner.

Pathologies in Neglecting the Future

The causes of neglecting the search for future consequences has many reasons and a partial inventory is offered below:

• *The reward structure in politics.*

Personal political survival must be assumed a major factor in the choices made in a decision-making situation. In the quest for voter esteem, those in power are bound to have significant regard for re-election schedules and probabilities when selecting which issues to make decision on and of the positions that need to be taken on those issues. Shorter-term issues are more likely to generate proof to the electors of the individuals success and capability than would a long run issue which would not see resolution until long after the election in question and maybe not until long after the death of the decision-maker themselves. It would appear that the only time a long term perspective is taken is when a politician is complaining about the shortsightedness of a member of the opposition party and this is also generally the only mechanism whereby any long term decisions are made, by a party locking themselves into a position by complaining about the other party. This can hardly be construed as a healthy balance between the long and short term considerations. This question of political survival has a match when it comes to survival tactics and this is the manifest preoccupation of politicians with the politics, the heckling, search for personal power, and prestige that can be gained from an issue rather than the careful and thoughtful analysis of the substance of the problem. This tendency is all the more dangerous when it is remembered that technologically-based policies are also increasingly political.

• *Pressures for rapid return on investments.*
The cost of borrowing money will always create an incentive for choosing those investments that offer a rapid return, as will uncertainty about the future. The sooner it is possible to cover the costs of an investment the safer the investment will be. This tendency is unpredictable, however, because interest rates change and so does the general economic outlook, but it can be stated that the attractions of rapid return investments are greater to late return ones.

• *The reward structure in industry.*
In the private sector the pressures are just as focused on the short term with demands for rapid and immediate evidence of success, although here the evidence of success is defined in economic rather than social terms.
Executives performances are measured by the quarterly corporate balance sheets, by indicators of corporate performance on the stock exchange, and by financial journalists reporting in the financial press. Within an organization, promotions and prestige is based on individual accomplishments subject to evaluation at frequent intervals.
The ruler for measuring success in the private sector appears to be the same as that employed in the political sphere, a benefit is gained for short run success that can be reported during these frequent evaluation intervals. In neither arena are we motivated to consider the future

because we are not yet there to worry about it; any success that accrues too far in the future would likely bring reward, power and prestige to a successor.

• *Fear of uncertainty.*
A fourth element is frustration because the future is always a fog of uncertainty, uncertainty as to ends and means. As Walter Lippmann (1966) stated, energies of our society are soaked up in the struggle simply to survive amidst uncertainty. Leadership does not have "the ambition to participate in history and to shape the future. Modern men are predominantly isolationists. They are preoccupied with the more immediate things which may help or hurt them. They are marked by a vast indifference to big issues and in this indifference there is a feeling that they are incompetent to do much about the big issues." Sir Geoffrey Vickers (1970) interpreted this pathological condition as being trapped by a state of mind: the past is no longer a confident guide to the future, and despite the loss in validity of old assumptions, there is great anxiety and little inclination to extend learning beyond what is widely termed linear thinking.

In a way, this loss in confidence is strange, because there is an increased understanding of the technologies, that we more and more employ, and that this technology is a set of systems that are generally easy to predict the effect of, in a technological sense.[26] Where the real problems arise are when we confront complex social behaviours, it is then that our confidence to forecast consequences is diminished.

These is another problem that arises from the complexities that the science-technology-society interactions create. Research is a mechanism for reducing uncertainty in a particular narrow field, however, the increased knowledge on a large scale, that this research makes available, increases choices and this increase in choices creates an increase in the levels of uncertainty.

• *Frustration with complexity.*
Complexity of social systems and processes adds another element to the feelings of confusion. Complexity, as earlier defined, results from the large number of components in technological systems, their functional and cultural diversity, their hyper-interconnectedness, and from changes in all these factors. Sensing these complications

---

[26] We know that if we turn on the television that we will get a picture and sound, if you push on the gas pedal the engine will speed up, etc. These are trivial examples of the predictability of technology however the examples are valid. Our societies are more and more comfortable with the technologies and are less and less confused by the technical consequences of the technology, where we start to become unsure are the social impacts.

individuals very rapidly come to understand that only their short-run behavior is relatively independent of the maze of constraints within which they are enmeshed and which limit their ability to make long term decisions which have a very low probability of success. Given the historical difficulties of mapping systems and of forecasting the long-run situation, even without the impacts of any decisions being made and the recognition that it was pure fantasy to further attempt to determine the effects of decisions on these unmappable systems it is understandable that decisions are made as they are; the longer-run problems are simply avoided. It is easy to understand why this was and continues to be so.

The observability of cause and effect relationships has disappeared as the systems have grown and increased in complexity. Unexpected consequences can be created by systems that are remote and appear to be unconnected to those impacted upon. Simple solutions continue to fail, even the ability to formulate and undertake corrective action is inhibited. Complexity can only be understood by the human mind - making connections, combinations and associations, but there are limits to human intellectual capacity to solve these compexities, and these also exists a threshold of problem-solving exhaustion (Rugge, 1975). No wonder simple ordered regularity has special appeal (Brewer, 1975). Finally, the presence of high levels of complexity can also mask individual responsibility and so removes the penalties for avoiding decision-making altogether.

• *Bliss by selected ignorance.*
How can we expect anyone to expend effort in confronting a threatening future, no matter how serious, if no means for reducing that threat are offered along with the warning.

This condition is made worse by the increasing rate of change and the unending and remorseless arrival of new problems that demand attention even before earlier arrivals have been adequately solved. There seems to be insufficient time, to think, let alone to solve.

• *The scarcity of time.*
There is no scarcity of information, it is possible to, literally, drown oneself in data and information at the flick of the proverbial switch. Where the problem may lie is not with the question of quantity but rather quality. We are suffering from excessive quantity and inadequate quality. A second problem is whether we have the required available time to search for and interpret the data made available. Phone, fax, telex, photo-copying, modems, compact discs and massive database technologies clog information and communication channels and distract people from a selection of meaningful priorities to deal with long-term strategies as well as day-to-day tactics.

As the complexity of decision-making continues to increase, and the allowable margins of error shrink, there should exist significant incentives for the investment of more time in data gathering and information analysis. The exact opposite is however true, the oppression created by more numerous and more frequent decisions and the hectic pace in most decision-making environments reduces the time available to contemplate, in any meaningful manner, the longer-term effects. At the same time, there is very little multi-disciplinary extended to the coordination of data analysis and information evaluation with desirable outcomes, prepared by independent and unbiased analysts who have no personal stake in the particular outcome in question.

• *Impediments to multi-valued goal-setting.*
While our society happily, loudly, and unthinkingly continues to make more and more demands, objectives cannot be listed and then ticked off one at a time as they are dealt with.

Not all demands that are made can have compatible solutions, and therefore proposed solutions will have undesired consequences on certain parties. What happens then is that those who can possibly be adversely affected by any such solution demand responses to, and a hearing of, their misgivings and fears. This response causes situations that force decision-makers to use up more of that rare commodity, time.

Therefore any attempt to tackle all goals simultaneously, and in anything like an optimal manner, surrenders to the squeaky wheel syndrome, those who yell the loudest get heard. And the long-range objectives, goals, dangers, problems and consequences, that rarely have a cadre of energetic advocates, are ignored for an array hat, at the time, are eminently justifiable however disasterous this may be in the longer term.

• *Pressures to reduce conflict.*
The social feedback system that reports on the effectiveness of decision-making initiatives, so admired in a democracy, by those who believe in democracy, is often reduced to an adversarial contests. With these conditions of escalating conflict, the decision-makers get jittery and understandably tend to ease tension by acting on those issues whose outcome may reduce tension most quickly and those issues are generally short-term in nature. Strange as it may seem, it is nearly always active citizen militancy that demands consideration of the longer view, an anomaly we will comment upon.

A key role of the decision-maker is to integrate these differences and generate solutions that are acceptable to all parties. At the same time, reconciling legitimate differences simply for the sake of compromise and the easing of tension and conflict may obscure the deeper perils. This practice of continually seeking compromise has become such an integral part of the decision-making process that it has been labelled by political scientists as 'partisan mutual adjustment'.[27] In the push for these compromise strategies, there is a lack of consideration of the longer-run consequences, especially if those who are party to the conflict argue only from their own short-term objectives and if immediate relief of conflict, whatever the longer-run penalties, is the prime motivating force.

One major exception deserves mention: the case of social intervenors (eg Greenpeace, Men of the Trees, Sierra Club, etc) who press for a concerted consideration of the future and of potential adverse impacts. The project delays, tough political bargaining and judicial processes the result from this intervention represents a new level of cost imposed by the public participation process. This increased cost burden on institutions within our society, that up to now have had no demands placed upon them to look into the future, is forcing a long-range perspective even simply as a defensive measure. That a concern for the future has to be forced upon certain players in our society by the existence of political conflict may itself be a manifestation of failure of the complete process.

Governments have traditionally attempted to deal with this problem of conflict by passing more laws. By the application of these laws, all the players could be expected to know and play the same game. But the major problem with the law is that it simply codifies already established social behavior and generally lacks the ability to sound warning signals and has none of the intrinsic functions required of an early warning system. If employed in isolation, as it often is, the law may only reduce the levels of flexibility we require to deal with the new situation and problems that confront us in an ever changing environment. The talent that all legislators have to earn brownie (good-measure) points on the basis of laws that bear their names as primary authors may paralyze the system with constraints and conflicts and new and useless laws that achieve very little, in either the short or long term.

• *Media pressure for the quick fix.*

Pressure for a politician to discount the future is further increased by increased demands for public accountability and by the increased levels of technology that allow for instantaneous notification to the public of decisions and recommendations with which they may

[27] The political situation in Croatia, Serbia, and Bosnia-Hertzogovina and the inability of the UN and world governments to solve the problem was a prime example of this failing.

disagree. Media exposure demands that decision-makers appear to be decisive and confident at all times, that they demonstrate an ability to instantaneously grasp the realities of any situation and make pronouncements and decisions independently of any outside advise, this leads again to decisions that can be no more than superficial solutions to the real problems.

Crises are paraded every day on our televisions for wide inspection by the population and that same population expects the implementation of rapid and simple answers. Whatever desire our cultures have developed for the 'quick fix' is further catered to by promises on the solution to short-run problems. This situation challenges the politicians will and desire to deal with complex, deep-seated issues with appropriately complex, and perhaps slow-acting remedies. To do so imposes levels of personal and political risks that few decision-makers seem willing to bear.

Remember that they were voted in to satisfy the wishes of their constituents and if their constituents want a fast and simple fix then its their job to give them just that. If their constituents don't care about the long term future then who are they to go against their constituents wishes? This is the most common defense that is used by a politician that is cornered about the question of long term decision-making failures.

• *Concealment of past error.*

A long view of the future may also expose the errors of decisions made in the past for reasons of pure expediency. Policy-makers are reluctant to admit past mistakes, change course or stop misguided programs. The bureaucracy that continues through changes in political leadership is especially sensitive to any such revelations. It comes as no small wonder that recommendations to solve, even the most dangerous and catastrophic, problems tend to be 'piecemeal, provisional, parochial, uncoordinated, unsubstantial, and lacking in prophetic moral vision' (Wagar, 1971).

• *Bureaucratic resistance to change.*

All organizations, both inside and outside of government, are initially created to propound a new idea, system or product, these enterprises, at some point in time, lose their vitality. At some point in time, the institutional comes to dependent not on its capacity to sense the significance of environmental change and to be a party to that change, but rather on the rate at which the organization itself can, or is willing to, change internally.

Vested interests dominate the organizations goals and objectives. When being attacked, as at some point in time these institutions always are, they continue to survive by retaining the loyalty of their adherents and in combating forces counter to their unvoiced interests and beliefs. They continue uncritically in pursuit of their originally stated mission, no matter how pointless it has become with the passage of time and events.[28] According to Drucker (1969), the inability to stop doing anything is the central degenerative disease of government.

Given their single minded pursuit of self-perpetuation, their lack of capacity for self-criticism, and their well-entrenched thought processes, the idea of change is always perceived as a threat and is therefore vigorously resisted. As Allison (1971) said, the bureaucracy does best tomorrow what it did yesterday. As one would expect, the more successful an institution has been in the past, the greater its ability to remain self deluded and to continue with its rejection of new ideas. Top level decision-makers feel guilty of the costs, resources and energies required to change the directions of their agencies.

One should be sympathetic, however, to this organizational inertia. Sincere attempts have been made at long-range planning. But in the annual budgeting process the future is heavily discounted also. All too often, when the budget is trimmed, it is the new proposals that are the first elements to be sacrificed. They collect together in filing cabinets for want of the necessary resources. Despite the promise of wealth for all from the application of high-technology, the availability of resources to encourage new processes is nonexistent, and we cannot expect the old and entrenched interests to be kind to the new kids on the block.

Why should we expect individuals to change behavioural patterns that have proved to be successful in the past. Talk to the CEO of any organization and tell him/her that the skills that got them to the top are no longer of value and that he/she needs to change their behavioural patterns. It is highly improbable that they will change. There is direct, personal evidence that the skills and behaviours that this person has historically manifest are of value since these are the traits that gained this person the career advancement that saw them into the Bosses Seat. All you can offer is unrelated evidence, evidence that is concerned with other people and other organizations - nothing to do with this person, in this situation, in this organization. You are fighting a losing battle.

The only way to win is to wait for the system to fail and for the individual to fail and then offer an alternative. Even then, more often than not, they will not change believing that any failure is only temporary and an aberration. Like Pavlov's dogs they have become conditioned to reacting and responding in a given manner and cannot

---

[28] It may however take decades or even centuries before these organisations are confronted because they are no longer of any value to society and in fact can start to become troublesome and even obstructionist.

now change. They have become predictable and cannot display the flexibility that is required in an ever changing world.

One last point needs to be mentioned in regard to continuous self evaluation on the correctness of past decisions, the need to institute corrective action, as well as to meet any new problems. Change is, by definition, an integral part of the self-evaluation process, yet this is anathema to most constituents of self-protecting agencies. Of the monies spent annually by the government for research and development, less than 0.1 per cent is devoted to analysis of the effectiveness of the other 99.9 per cent.

There are a plethora of programs, agencies, systems, processes, procedures, and regulations that could, without harm to the country and its people, be eliminated almost overnight. One of the major reasons that they will continue to exist is that there is insufficient time for any decision-maker, with the required levels of power and authority, to gain an understanding of exactly what can and cannot be eliminated. The use of advisors and consultants in this arena is of little help since they are generally stonewalled, and very effectively, by the organizations they have been asked to review and by other agencies that are asked for comment on the effectiveness of the agency in question. There exists an unwritten understanding that agencies do not negatively comment, other than in battles for power, on each others performance or value.

• *Risk rather than crisis avoidance.*
At any level, self-generated change carries the risk of creating a destabilising situation. Things can get worse as well as better. When the forecasts of crises are too ambiguous and uncertain, and when in the midst of complexity there is doubt about how, when and where to apply corrective action, latent tendencies to avoid risk that are present in all institutions are dramatically reinforced. Given the intrinsic difficulties of striking bargains even in static situations, that requirement is even more demanding in the presence of change. The higher the stress on the decision-makers, the greater can be their desire for spontaneous self correction and with the knowledge that they are powerless to, in any coherent and meaningful way, motivate change they can, understandably, metaphorically crawl under the blankets and wait for it all to blow over.

• *The failure of strategic planning.*
In addition, confidence in long term planning has diminished, partly because projected events often do not happen; the public is fed up with crying wolf over propagandized threats, plans are shaped around some fictional static and homogeneous average that represents individuals in our society, and the public have come to reject 'top

down', master-minded planning as being far too dictatorial and not matching the ideals or needs of today's communities. Also, the remedial measures proposed and implemented are often counterproductive; continuing to add highway lanes does not relieve city traffic congestion, having more police does not reduce the crime rate, having more schools and universities does not produce a smarter more adequate populace.

There still exists a serious conceptual gap between planners and those who the planners are supposedly planning for.

• *No liquid resources.*

There is also the lack of necessary resources that are required for looking into the future. To consider and create change requires the application and availability of the right types of people and also money. The information and analytical capabilities that are required for the development of a comprehensive and capable early warning system appears to be generally unavailable in the numbers required to ensure our collective safety. Contingencies are not well considered during the competitive rush for resources during periods of its allocation. Despite the multiplier effect attributed to technology, and the implied reductions in resource requirements, no reserves have been collected together in the system, any resources released by the application of technology are absorbed by the never ending demands for instantaneous gratification. Without the necessary resources, incentives to examine available options diminish because they become abstract exercises in fantasy.

• *Barriers in culture.*

Politicians understand the problem of discounting the future. By and large, they want to embed it in their decision-making processes, but they cannot. It is this inability to embed their understanding into the decision-making process that causes their condition to be pathological. Given these numerous manifestations of uncertainty, of complexity, of scarcities, of Babel and of impotence, it may be expecting too much of political leaders to get out in front of the voters, who are themselves buffeted with day-to-day crises, frustrated, demonstrate an appetite for instant gratification, and lack awareness of 'the situation' and offer a cohesive and coherent vision.

The credit card economy so much an integral and unthought of part of our culture is yet another manifestation of imperatives toward immediate satisfaction. Thus, we may have to face up to the insurmountable barrier of public apathy and indifference, maybe even active and violent hostility, to the long run, stemming from cultural as well as private and personal attitudes to self needs, wants, desires and greed.

The lack of a common understanding that the future begins today is one more source of avoidance in the policy making apparatus to signals about the future. Citizens' protestations as to the catastrophic futures they do not want are so weak as to be lost in the roar of demands for immediate action to alleviate today's discomforts; and the policy-makers are not listening to the whispers, even their own, but only hear the roars from the mass of the public, quite understandably.

These pathologies of the short run are both a source and consequence of stress in decision-making. To understand the implications, it is necessary to now understand the operational issue of how these pathological behaviours affect the performance of policy decision-making.

Ranking Policy Outcomes

Social outcomes of policy can be derived from three sets of criteria, in terms of impacts on the structure of the system, on the functioning of the system, and on value preferences of the society that will be affected by the system. While social, economic and environmental effects are included, the political and psycho-cultural impacts have been deliberately highlighted. For purposes of explanation, these impact factors can be transformed into questions about consequences and can also be used as a check-off list for decision-makers.

Political Impacts posed as Questions

1. How is the decision-maker's stature altered?
2. Will conflicts in goals of different interest groups be polarized or reconciled?
3. Who wins and who loses and by how much?
4. Will existing institutional roles and behavior be altered?
5. Are new policies consistent with existing policies, social norms and laws?
6. Will new programs be implemented efficiently and effectively?
7. Will public confidence be affected in the ability of decision-makers to set goals, allocate resources and reduce conflict?
8. Will the determination of top decision-makers to exercise legitimate power in the face of obstacles be altered?
9. What future effects are conceivably different from immediate effects; who will be affected and how?
10. Will constitutional guarantees and practices affecting individual obligations and rights be changed?
11. Will maintenance of law, order and safety be altered?
12. Will there be changes in influence, by citizens, on policies and their generation?

13. What will the impacts be on the natural environment and on utilization of non-renewable resources?
14. Will the exercise of power be more or less centralized, with changes in local, state and federal authority?
15. Will equity or opportunity for individual self-expression change?
16. Will the capability to transmit undistorted information be altered with respect to knowledge generation through research, to analysis and to dissemination?
17. Will our capability for early warning of peril and for contingency planning be changed?
18. Will organizations levels of publicly accountability alter?
19. Will future opportunities or evolution of social goals be blocked?
20. Will the entire system be better able to cope with surprise, be more resilient and ensure the availability of resources?
21. What will the effects be on open-mindedness to future change or rate of change?
22. Will there be a change in our capacity to image the future?
23. What changes may occur in overall cultural patterns?

These have been put into order according to the swiftness of expected repercussions. These gestation periods of impact may range from weeks to decades. Clearly, the delay in perceiving effects varies enormously from one case to another, so that the rank order is necessarily an abstract average. These potential impacts are next ranked in the second and third columns of Table 3.1 by the importance accorded each element by a hypothetical policy official and hypothetical citizen. In so doing, we are not here concerned with the intensity of each impact, just how urgent each impacted party considers it is to investigate. The ranking is thus a technique of identifying what for each party are the 'right questions'. As part of their cogitation, if a decision-maker or impacted party can develop corresponding answers as to what could happen, a comparison may then be made with what that person thinks should happen. This constitutes a pre-crisis decision assessment.

A major difficulty arises when one attempts to order these impacts in any meaningful way. Their importance will vary from one situation to another either because the issues are different or because there exists a state of crisis because of some other decision has been made that had unexpected adverse impacts. There is also the difficulties associated with attempts to generate averages of divergent views of either the policy-makers or citizens; the average (if there is any such point in reality) may not be a simple arithmetic averaging or the isolation of some mid-point between extreme positions. But even after advertising these stumbling-blocks, there is revealed a pattern of conflict between short- and long-term values when comparing the questions and problems that excite the citizen with those that animate the politician.

Table 3.1 Ranking of Impacts (Please see concluding note on the final page of this book)

	Time to impact	Policy maker	Citizen	Society
Continuity in political power of policy-maker	1	1	19	23
Reduction (or increase) in social conflict among interest groups	2	2	5	18
Distribute outcomes of decisions	3	3	2	19
Continuity of institutional structures, public and private	4	4	12	21
Compatibility with existing laws	5	9	17	13
Economic efficiency (lack of waste) in implementation	6	6	3	12
Continued ability of government to govern	7	7	14	14
Maintenance of political will in Leadership	8	5	13	22
Changes in governmental role in relation to private sector	9	8	4	15
Indirect and future benefits, reductions of peril risk and costs	10	11	11	2
Maintenance of social cohesion, order and freedom from violence	11	15	1	6
Access of citizens to political process	12	17	6	10
Protection of environment and conservation of natural resources	13	16	8	7
Trends toward state and local authority	14	10	16	16
Changes in quality of life, opportunity for self-expression, equity	15	18	7	11
Strength of information, scientific research and monitoring	16	12	10	4
Capacity to appreciate the situation and contingency planning	17	14	20	3
Accountability and public information on government performance	18	19	11	9
Preservation of future options	19	13	15	5
Capability of entire system to cope with surprise (resilience)	20	20	23	8
Attitude (hospitality) to future change	21	22	18	20

Capacity to image the future	22	21	22	1
Fundamental changes in cultural framework	23	23	21	17

In the struggle for power, citizens place high importance on their access to the those in power, by exercising the opportunity to actively participate in the decision-making process. Politicians, on the other hand, are far more concerned with expanding their personal influence and power, with maintaining friendly relationships with key institutions in our society on whom they depend for implementation of policy and with organizations which display a similar appetite for influence. Citizens are also, in many cases, the leaders in defining shifts of social priorities and value systems associated with the quality of life.[29]

The fourth column is yet another ranking of impacts, now taking as priority those consequences that influence the long-term health and survival of a society. This might be thought of as a ranking, given today, now by a citizen who would not be born until the year 2000. Somewhat contrary to what might be expected, in focusing on the longer term, the order is not the exact opposite of the first column, although there does appear to be a general reversal in the pattern.

If we now comparing the four columns, we find that neither the policy-maker nor the citizen are seen to consider the long term, 'iffy' questions as being of major significance. No matter how committed we, personally, may be to the future and its consideration as a serious undertaking, it would be simple minded of us to expect a meaningful trade off of the present for the future. Therefore the issue is one of attempting to maintain a level of balance.

One ranking in the table, that on 'capacity to image the future', particularly the futures we do not want, is so unnatural as to demand an explanation. The importance of this social characteristic has been underscored by numerous social philosophers. Fred Polak (1973) puts the case this way:

> The image of the future can act not only as a barometer, but as a regulative mechanism which alternatively opens and shuts the dampers on the mighty blast furnace of culture. It not only indicates alternative choices and possibilities, but actively

[29] Laws relating to a large number of issues in our society were first voiced by the public, leading the decisions makers rather than having the decision makers being self aware by a concerted effort to read the wind and monitor the environment. Decision makers are too often surprised by the strident demands of the populace which, if they were performing their jobs in a satisfactory manner, they would not be.

> promotes certain choices and in effect puts them to work in determining the future. A close examination of prevailing images, then, puts us in a position to forecast the probable future.

The validity of this contention is supported by some historical evidence. At one pole were the Dark Ages, seemingly without images. At the other, as Dostoyevsky has suggested in his 'Diary of a Writer', an ethical image of the future has always preceded the birth of a nation. The importance of image development is also stressed by Ellyard (1991) who proposes that a preferred future cannot be achieved without the ability to image that future. Ellyard states:

> The creation of images and visions of what such a world could look like is a critical part of helping to realize it.

in discussion of the world of 2020. He goes on to say:

> A 'vision' then becomes a 'prophecy' which can be fulfilled by coordinated effort ...

Imaging the future does not mean forecasts or prophecy, nor, necessarily, the adoption of a single vision. The capacity to image may possibly be described as the capacity for fantasy, so easily evidenced in children, perpetuated in myth and folklore, and as observed in the creative activity of adults in play and in the arts. There is a danger though. There is the possibility that newly manifest cultural trends that have fastening on existentialism supported by our new technological artifacts may have suppressed or maimed these tendencies. Indeed, there is a question as to whether new social trends among the influential elite not to have children may influence their disposition to consider a future that not only excludes themselves and their descendants but also those of others.

Needless to say, the rankings by the authors is totally subjective, totally lacking of any evidence that could possibly imply verification, and, in column four, reveals simply their own set of values. No proof is claimed, or required, for the table to support the assertions made. Readers might wish to replicate the table and fill it in based upon their understanding of the priorities of the various players, we do not believe it would be much different. The exact orderings in the various columns are not vital to the arguments being proposed.

The point that needs to be made is, that under conditions of stress and the pathologies of neglecting the future, there seems to be an inverse correlation between the importance accorded these performance criteria by decision-makers and the time until any impact becomes manifest. That is to say, at the moment when a choice must be made, consideration is first given to those questions which have immediate known consequences, with diminished attention, if not complete disregard, to those with delayed and usually less certain consequences.

It is very dangerous to suggest that all decisions follow the patterns just outlined or to believe that neglect of the long- for short-term considerations is always costly or dangerous in the long run or to suggest that all of the listed performance criteria are deserving of equal attention.

In the most cynical interpretation of this list, a choice might be made only on the basis of political expediency by an American President, for example, close to re-election, in trouble with the Party, the electorate and business. The bureaucracy and committees are known, from research findings, to push for legislation from ambition to maintain a constituency or to enlarge their influence.

There is another side to expediency that needs to be mentioned, because the term is now considered to be derisive. Expediency in the sense of a politician's acting on the basis of self-interest to retain popularity with an electorate also reflects behaviour we all expect from those we elected in a representative democracy. Voters would rightly object to indifference to their concerns and preferences of arrogant representation and representatives. This mechanical, uncritical responsiveness to the demands of an electorate, however, begs the question of the role of a policy-maker as teacher, to move a constituency from being narrow minded, parochial, isolationist and uninformed to being better informed and having the ability to make broader considered judgments.

Next comes the question of what constitutes expediency? With the increasingly divergent desires of constituencies and the general lack of a perceptible single overriding consensus on any issue, pressures arrive from many directions. Each decision made by the decision-maker is found to have fault and is closely followed by the complaints of a group of dissenters, often claiming that the representative caved in to pressures from an opposing pressure lobby. No matter which way a decision is cast, the decision-maker will always step on some group's toes.

Under these conditions, how is it possible to made reasoned and reasonable decisions for the benefit of all?

Let's return to the list of impacts, the second half is of interest not only in terms of delayed impacts but also helps us to define the health and flexibility of society. That is, it is the delayed measures of social performance that also have to do not only with the correctness of a decision but also with the future abilities of the decision-making system itself to work correctly.

Thus the neglect of the future may not only invalidate a particular decision; it may undermine our future ability to decide. That is, all the different types of decision-making failure described are continually reinforced and are likely to reduce our future ability to make meaningful decisions.

We have taken a position on dangerous and, what we believe are, significant trends in the nature of the arising threats. However, we should also consider new trends in decision-making practice so as to develop a balanced consideration of these long-range forebodings.

Two contradictory trends are becoming obvious. The first is that decisions are getting more political, with very obvious incentives to concentrate effort on the short-term. At the same time we have more in number and more effective voices in our society calling for a shifting of effort to achieve a more balanced understanding of the consequences of neglect of our future. It is also possible to distinguish changes in social priorities that are now represented in the political sphere. These trends are not consistent or uniform but change rapidly in response to crisis and pressure.

We need therefore to discover if long-term awareness is growing as rapidly as are the global threats and as rapidly as is the inclination to act for expediency.

In discovering mechanisms that will help us to improve decision-making and to isolate targets upon which to concentrate effort and resources, it would be a disaster to examine only the obvious and manifest dysfunctionings, there is far more, not obvious but hidden below the surface, that must also be examined, in detail, before we can feel secure in our knowledge of the workings of the decision-making system. Although policy is pronounced from high-level fraternities a study of these brotherhoods is insufficient for the task. All actors in the system are an integral part of the decision process, and we need to unravel the mechanics of this system to see how it works. This implies a study of all levels and actors.

4 Reasons for Failure in Planning

What, sir, you would make a ship sail against the wind and currents by lighting a bonfire under her decks? I pray you excuse me. I have no time to listen to such nonsense.

Napoleon Bonaparte, 1803 in discussions with Robert Fulton, an Engineer, concerning steam powered ships.

Men might as well project a voyage to the moon as attempt to employ steam across the stormy North Atlantic.

Address to the British Association, 1838, quoted in Patrick Moore's *Space in the Sixties*.

Atomic energy might be as good as our present day explosives, but it is unlikely to produce anything very much more dangerous.

Winston Churchill, 1939.

Planners

Knowledge of the Environment

Comment on the size of the environment of interest (massive in any real problem), the inability of any person to have complete knowledge of that environment (planners too rarely seek knowledge from others that is needed to make a rational decision) and even if others are referenced as sources of knowledge there is still the lack of ability of any one person, the planner, to make any real sense of all the information and data that can be supplied to given an understanding of the environment.

Understanding the Complexities

Even if the planner does seek the knowledge of others it is virtually impossible to disentangle all the complexities of the problem and plot a course that will bring about the desired end-effects. The problem can be likened to a rubik's cube with 1000's of faces not just 54.

Understanding the Inter-Connections

There is also the problem of interactions between factors. Too often planners blithely believe that they have a good handle on the problem. "If we increase speed limits we will increase road deaths. Road deaths are a bad thing; so don't increase road speeds." The analyses are often limited to this example of first order planning.[30] The dangers in applying only first or second order planning is that other potential side effects of any decision are disregarded and it is certainly possible that some of the side-effects are of very great danger to humans, as individuals,[31] societies and humanity as a whole.

Planning Tools

There is a total lack of tools to assist planners in the actual actions required of planning and decision-making. For too long has there been a belief that good planners are born and not made. In some regard we agree. An individual needs to have the capability of abstracting complex structures and viewing a problem much like a set of forces, each acting on others and on themselves. Pulling this way and pushing that, the planner needs to be able to visualize these forces and resolve the resultant/s.

Some people have this capacity inherently, others need to develop it over time (this can be an expensive process if too many mistakes are made by the planner in the process of learning). What will help all planners is a tool to assist in making decisions. If a tool that displayed the forces (factors comprising the environment), which other forces they act upon and to what degree (inter-factor effects) and displayed

---

[30]The order of a problem can be demonstrated as so:

1st order	If A then B
2nd order	If A then B and if B then C
3rd order	If A then B and if B then C and if C then D, etc.

[31]A prime example would be the limited testing of consumer products. A new electrical system is installed into a car. No need for testing since the car company knows all about car electrical systems. Problem arises after a number of cars have had rear-end accidents and have caught fire and killed a number of people. First order planning has proved itself to be insufficient to the needs of the company and its customers.

the resultant of all these forces and their effects; then we have a useful tool to assist the planner. No such tool currently exists, but we are in the process of constructing one.

Political Leaders

Political 'Realities'

"If a decision is going to cost me votes I won't make that decision; independent of how right it may be."

A political 'reality' is simply another word for self survival. If a particular decision will cost a politician votes then political 'realities' require that that decision not be made or that the decision be made in such a way as to not cost votes. The politicians justification for these 'honesty' lapses in decision-making is that if the decision were to cost votes then it is obviously not a decision that the electorate wants and since the politician was placed in power to fulfil the wishes of the electorate to make a decision that would cost votes is patently against the wishes of the electorate and is not a decision that should be made by any 'honest' politician.

Short-Term Focus

"Every decision I make must pay off between now and the next election". No politician is guaranteed a future in politics and as such it is necessary that each political appointee attempt to ensure their future by pleasing the electorate. This implies the making of 'brownie point', 'having runs on the board' or in other ways showing the electorate the progressiveness, dynamism, astuteness, honesty, intelligence, concern, etc., etc., at the earliest stages of their political careers. There is also a need to continually reinforce these perceptions in the electorate and hence much of a politicians effort is aimed towards furthering their careers, not necessarily for the benefit of those supposedly represented.

The worst abuse of this system is as the time for an election grows closer. The need for a politician to be at the forefront of each electors mind is seen as the most important single factor in existence. It is possible, at times like this, that a politician will make decisions and promises that are purely aimed at pleasing the immediate needs of the electorate. The consequences of such actions can and have proved to be disastrous later when the promises have to be kept or the price paid. Far too often the price is not paid by those who benefit and this can be exampled by the existing degradation of the environment. But it is others; those who have not benefited who pay the price to clean up the mess – it is the living and those still to come who pay the price. The same abuses are made by politicians far too often and it is not them who pay the price of their political greed but it is rather you and I, our children and their children who will suffer from the incompetence and ignorance manifest in their behaviour.

The easiest way to 'prove' that a politician has all the qualities listed above is to agree to satisfy their wants. If an individual says to you, "I want less crime" and you agree with them and then push for the death penalty for serious offenders, you are not seen as a fool or a danger to society, rather you are seen as a person with astuteness, concern, dynamism, individuality, in fact with all the attributes that the requestor believes themselves to have and given the average level of egoism in the world that is almost every virtue known to mankind.

There is another danger to long-term planning in politics and that is the fact of changing government. It is eminently possible for a politician to start a project that will require ten years to reach fulfillment to then loose power after eight years and for the new government to take the credit for the project when it does finally pay-off. This leads to a very cynical observation, that given there is a high probability that your party will loose power in the next election there is a high level of motivation to put in-train projects that will pay-off during the new governments term of office; however, the pay-offs to these projects would be negative rather than positive. It is then eminently possible, from the safety of the opposition, to point an accusing finger at the new government. If however; your party retains power you are in the position of being totally aware of the consequences of what you have put in operation and are in the best position to derail, stop or limit the negative effects of the project.

We find it a continual surprise that the population can so easily be fooled by this simple tactic. It appears as if all things that happen, for good or bad, can be laid at the feet of any government if the event occurs during their term in office. If on the first day in office for a particular party was the day a national strike started that had been fomenting for the last year, then the newly elected party would be seen as responsible for the coming about of the strike not just for fixing it up (this may be a bit of an exaggeration but not much of one, we believe).

The Media

Often it is easy to believe that comments like the following are being made behind closed doors;"If a decision causes too much controversy then I won't make a decision. Better No Press than Bad Press".

The Public

Also comments like the following are often heard;"I was put in power to fulfil the wishes of my electorate." Let's be honest a politician only ever says that when it's a decision they want to make anyway.

Why fulfil the wishes of the electorate if they are not aware of the consequences of a decision they are attempting to force through? The consequences of the wishes of the electorate, given they rarely have all the information, will bring about far greater dangers, evils, etc., than they are attempting to clean up. A prime example is in the jailing of juvenile offenders. The argument is an old and long one, should juveniles accused of relatively trivial crimes be jailed with hardened criminals? I think not but there are a large number of the public who believe they should be. Should a politician strengthen the juvenile justice laws at the request of the public or should the laws not be tightened up so as to avoid the worse situation of having hardened juvenile criminals coming out of our jails once they have been trained in prison by the hardened criminals?

Who proved that one hundred idiots were smarter than one intelligent person? This is the basic tenant of a democracy. We, as a democratic society, say quite unashamedly that as soon as a person reaches voting age that they are capable of assisting in deciding on the future of their country.[32] This appears to be a nonsense. There are persons who cannot read and therefore cannot truly become aware of the issues, there are those who are over the legal voting age but who's mental age is well below, there are those who are criminals but who have simply evaded capture or conviction, there are those who are mentally unstable and sociopaths and psychopaths abound in our modern civilized societies and we trust this random, unruly, ignorant, criminal, unbalanced mob to elect a government? You may, we don't.

Toeing the 'Party Line'

It is not possible to voice a dissenting vote. Each politician must "toe the party line", independent of their own conscience. If any politician fails to mimic their peers then they are ostracized.

Level of General Education

Politicians, in Australia, are often ex-farmers or trade-union organizers. The first group want us all to work eighteen hours-a-day seven days a week, sleep for five hours and pray for the last hour,

---

[32]In some countries you are not allowed to vote if you are being held in a prison or mental institution.

survive by eating sheep droppings and live in Corn Flakes packets, the second group wants us all to 'earn more than the average income' (a statement made by an Australian Labor Politician. Note: that it is impossible for every earner to earn more than the average). Want us all to have two or three cars, own our five bedroom, two bathroom, two garage homes, have a free university education and get it all by not working at all.

Understanding the complexities of running a country is, for any average person, we would contend, a daunting task. Why is it then that a small cadre of people appear not to be daunted, is it that they are so sure of themselves and their capabilities that they actually believe that they can perform such a task and that they have the intelligence, levels of knowledge, and understanding to undertake such a task successfully?

What generally saves these people from total failure is the existence of a group of people who survive each government and continue to work within the political system — public servants. These people can maintain the continuity that is required, they can act as sounding boards for new governments - especially those that have not been in government for a number of years.

Political Advisory Groups

Politicians have massive departments to support them, to assist them when needed, to advise on tactics and strategies; but still they have large bodies of advisors. Why? The answer is relatively simple. The advisors are worried about the politician, the public servants (if they are not political appointees) are concerned with the public. Often these two groups do not have the same objectives and can have significantly divergent views. It is patently possible, and probably occurs far more often than we would like to believe, that a politician makes a decision to ensure their self survival against the wishes and/or interests of their electorate, their state, their country or the world.

Politicians therefore often fail to believe advise given by public servants and rather trust their political advisors. These advisors are there to help the politician execute the party political line without getting too many electors off-side or at times to advise on actions that are patently detrimental to the electors if there is advantage to be gained politically (and sometimes financially) and the electors are not likely to become aware of the action or its consequences.

Industrial Leaders

Short-Term Perspectives

Need to continually obtain profits. No good working on long-term stuff if it will adversely affect profits in the short-term.

Most industries apply research performed at universities or other research bodies set up by governments. There is very little, if any, significant research being done by any but a few companies.

Share-Holders

Shareholders must be kept happy. This implies increased share prices and good dividends. Again a short-term perspective is vital for senior management.

Believing in Accountants

We believe that most industrial leaders who may have read this far will stop here and state that this book is all a waste of time, for us to say that accountants should not be listened too is tantamount to industrial suicide. We must disagree and feel that we can prove the fact.

In a manufacturing environment an accountant will state that the major objective is to minimize average product cost. Therefore we continually search for mechanisms to minimize set-up times, achieve high utilization of manpower and equipment, employ economic order quanties (EOQ), economic batch sizing (EBS),[33] product line balancing, etc., etc. How wrong can we be?

The object of a production facility is to contribute to profits. It is possible at times that any attempt to minimize average product costs, by attempting to minimize idle time and maximize utilization will cause delays farther inside the system and this lost time can never be regained. So the output of product falls and sales are lost. If we had allowed time for a man or machine to be idle, waiting for an important order to arrive so that processing could start immediately, then the product would have been in the warehouse to fill the order in a timely manner, as it is the order is lost and so is profit.

We lose money all because an accountant insisted that we must produce product at minimum cost to be as profitable as possible. What

---

[33]There is a whole area of management (called Management Science or Operational Research or Operations Research, depending upon where you come from) that is dedicated to the scientific analysis of business operations with the objective of making these businesses as efficient, profitable, effective as possible. EOQ, EBS and product line balancing are but three of the many tools used by the profession.

a load of bunkum! Is it better to sell one thousand at $100 profit each or to sell one hundred at $200 profit each? If you listen to your accountants they will create a system, within your company, that will cause your people to work for the one hundred at $200 profit each. Beware the accountants! (See Goldratt & Cox)

There can be profit in leaving staff or machines idle, in certain situations. Have a look at the diagram below: We have three machines, all of which process at the same average speed and have the same distribution of processing speeds. These machines can process anywhere from 1 to 6 item every minute (this value was chosen so that you can simulate the machines with a six-sided dice if you wish)[34] and the actual number that will be processed is not known until the minute is complete (in other words throw a dice, once for each machine for each minute, and the number showing is the number that will be processed by that machine for that minute). On average each machine should process 3.5 units per minute, this is the average of a six sided dice. However; in this fully balanced system (something all factories aim to achieve is to ensure that each machine be able to match, exactly, the output of all other machines in a particular production run) we will allow the first machine in the sequence to assume there is a product inventory sufficient for the machines needs. If a three is thrown for the first machine, on the dice, then three units are passed onto the next machine, if a one is thrown then one unit is passed onto the next machine. Once we have thrown the dice for the first machine, we move onto the second. We throw the dice for the second machine and pass on the number of units that match the number shown on the dice to the third machine. However; if the number showing on the dice is greater than the number of units available then you can only pass on the number of units available, as would happen in a real factory. You cannot pass on non-existent units. Let's try the simulation and see what happens. We will also ask ourselves a number of questions after the simulation run.

Given that we cannot speed up any of these machines, we can now ask the following questions.

Question 1. Should any of the machines be keep idle on purpose?
Question 2. If answer to Q1 is 'Yes' then which one/s should be kept idle?
Question 3. If answer to Q1 is 'Yes' then why should they be kept idle?

---

[34]This is, in general, how machines do work in a real factory. The only difference is that the variation is far lower, in a factory, than in this example. The variation of one-to-six was chosen to allow for a dice and to highlight the causes of variation.

and the answers are:

Q1. Yes
Q2. Machine 1 should either be slowed down or be kept idle at times.
Q3. Note the increase in work-in-process held up by machine 2 (a total of 19 units). If machine 1 could have reduced production by 19 units then we could have reduced work-in-process inventory by 19 units and saved the company money without reducing final output. Try the simulation yourself, using your own dice. The result you get will be practically the same. There is no escaping the obvious lesson, at times it is more profitable to purposefully slow down machines or to leave them idle.

Forget the accountants and their insistence on absolute minimal average unit cost, they will lead you astray by insisting that all personnel and all machines work all the time; leading to ever increasing work-in-process inventories. Do some financial calculations on the above example and work out a hypothetical average unit cost for the results above and if you had stopped production for 19 units worth of production. You will note that average unit cost is higher if you stop production, but output hasn't changed; only inventories have gone down and you know that will save you real money.

Table 4.1 Determination of Machine Productivity

	Machine 1		Machine 2			Machine 3		
Minute	Dice	Pass On	Inventory	Dice	Pass On	Inventory	Dice	Output
1	5	5	5	3	3	3	5	3
2	5	5	2+5	5	5	0+5	4	4
3	4	4	2+4	3	3	3+1	1	1
4	2	2	3+2	1	1	3+1	1	1
5	4	4	4+4	1	1	3+1	3	3
6	6	6	7+6	2	2	1+2	4	3
7	4	4	11+4	2	2	0+2	6	2
8	4	4	13+4	2	2	0+2	6	2
9	5	5	15+5	2	2	0+2	5	2
10	3	3	18+3	2	2	0+2	3	2
11	4	4	19+4	2	2	0+2	1	1
12	3	3	21+3	1	1	1+1	5	2
13	4	4	23+4	6	6	0+6	2	2
14	5	5	21+5	4	4	4+4	3	3

15	6	6	22+6	6	6	5+6	6	6
16	3	3	22+3	2	2	5+2	5	5
17	1	1	23+1	1	1	2+1	4	3
18	2	2	23+2	4	4	0+4	1	1
19	2	2	21+2	1	1	3+1	1	1
20	1	1	22+1	4	4	3+4	6	6
Totals	73	73	19	54	54	1	72	53

Narrow Focus

Companies look, far too often, at the functions/services that they offer today and see that as the full extent of their business. There is a need to view an organization holistically as part of a larger environment, searching for synergistic concordances and beneficial attachments. Too often businesses fail, after years of brilliant performance. The reasons often are that some other organization saw an opportunity the historically successful organization failed to see. This opportunity has seen the death of those businesses that failed to take advantage of a new direction, system, process, or product.

Environmental Niching

Most, if not all, industrial leaders are experts in only one type of environment. They can handle times when they are confronted by a 'Bear' market, or can handle periods of consolidation, or can handle periods of acquisition. However, it is a very rare corporate manager who can handle all of these situations and/or others as well.

What is found is that corporations are well managed for certain periods of their life and are then badly managed as the environment within which they work changes with the passage of time. There are a number of reasons why these organizations are left to bad managers in changing circumstances.

The company is owned and operated by the corporate manager under discussion, those responsible for changing the CEO fail to see the danger or believe that the current failures are only temporary (since the manager has proved himself in the past and the current situation must therefore only be temporary), the manager's golden parachute is of such a size that the company cannot afford to get rid of the individual concerned or there is a belief that the survival of the company is dependant upon this individual (ie the persons personal charisma is vital to maintaining or attracting new customers, the person is the only one who understands the complete business, etc.).

Religious Leaders

Lack of Concern for this World

If you are a non-believer, as most humans today are, you would be a fool to place your trust in an individual that cares more for your soul, in the hereafter, than in your current physical manifestation. You are worried about hunger, crime, disease and poverty. They are concerned about hell, purgatory and heaven. When the Pope, in 1992 as part of the World Summit in Rio, stated that population is not a direct cause of poverty he was possibly making a decision on souls not humans.

The larger the world population the more potential believers there are and the more potential souls in heaven. What religious leaders seem to fail to realize is that there are also more potential souls in hell and that there is a direct link between poverty and sin, as defined in the Bible, Talmud, Koran and all other religious tomes.

Dogma and Lack of Flexibility

One of the most damning traits of nearly all religions is that they are based upon dogma. The books of religion were written, in almost all cases, many centuries ago. The world and its people have changed and yet our spiritual needs are attempting to be satisfied by those who read and preach from words written to sooth the souls of people from centuries past.

What is the relevance today of a prohibition on the consumption of pork or shell-fish? Where is the sin, in any religious work, of polluting the rivers and seas, of contributing to acid rain, of being a banker who instigates a foreclosure on a struggling farmer? To maintain a sense of morality in an ever changing world, most have had to add to and embellish the messages of the worlds religious leaders. We have had to do this to ensure that we can distinguish between right and wrong, for if we simply accepted the narrow message and perspective of religious leaders then we would still be struggling in an environment that was a duplicate of the middle-ages.

Refusal to Consider Alternatives

If you are to remain an accepted member of most religious orders there are certain rules that you must obey. No different to a club that has rules and if you break or disregard these rules you will be asked to leave, so with these bodies. It is not possible to be a Christian and refuse to accept the divinity of Christ, or a Muslim and refuse to accept that Mohamed is the Prophet of Allah.

If you research and discover more about the various religions you will discover that the rules are far broader and penetrate far deeper into the day-to-day life of a true member of any religious sect. The more traditional and/or stringent the religion the more inflexible it becomes, the less able are its members to instigate change, any change; for better or for worse.

We believe that it was the more relaxed religions that created the impetus for the change that has seen the rise of the "developed countries". These religions were Lutherism and Protestantism with protestantism being the more powerful driving force. In the relatively relaxed atmosphere of the protestant church it was possible to propose new ideas that would have been damned as heretical under the more stringent mores of the more traditional religions.

New ideas created new industries, new ways of increasing the knowledge of the people, new methods, new techniques and slowly the emergence of new cultures. Why is it that in Europe it has generally been the non-orthodox religious countries that are wealthier and more industrialized. The countries that shifted to less stringent religions moved forwards the rest moved forward only on the tails of these others.

Genetic Diversity in Organizations and Structures

Politico-, Corporo-, Econo- and Societo-Genetic

Two cases serve to show the lack of politico-genetic diversity in politicians. Early in June 1992 Israel went to the polls to elect a leader for their next period of office. The incumbent, Yitshak Shamir lost at the polls with the message of 'We will not surrender any of the lands we have taken from the Palestinians.' his opponent won with the message, 'We will negotiate.' I remember seeing Shamir on the television news and the commentator mentioned that Shamir looked confused after he had lost the election.

That confusion was understandable. For years this person had been trained, just like one of Pavlov's dogs, to believe that if he reacted in a certain way, fight hard - don't give an inch - etc., then he would be lauded, respected, listened to and that he would retain power. Suddenly overnight, or so it would appear, the message no longer works. He pushed the right switches, as he had done before, but the reaction was now different. The environment had changed (The same happened to Bush in the 1992 televised debates with Clinton (We won't mention Perot's lack of contribution)).

No longer did the old words and statements conjure up visions of growth and a good life in the minds of Israeli voters but rather created visions of blood and death and loss of international respect. Shamir lacked the requisite politico-genetic diversity to deal with a changed environment.

It must be kept in mind that there is no assumption here that Rabin (the winner in the election) had the requisite politico-genetic structure required to survive; only time will tell if he does have what is needed. In many instances what we see is the simple taking of the opposite view. Rabin had to offer an alternative to Shamir to stand apart from him in the polls. In this case it worked. Alternatively, Rabin may have read the electorate and decided on the correct course of action or the message he gave was the message he believed and his politico-genetic structure was exactly what was needed in this new environment to succeed.

The second case that can be demonstrated is the resignation of Premier Griener of the state of New South Wales in Australia in the same month as the case described above (June 1992) in what became known as the Terry Metheral Affair, Premier Greiner offered a member of his political party a lucrative public service (government) position. Terry Metherill accepted the position and resigned from the political arena.

The behaviour displayed by both parties, Griener and Metherill, would have been acceptable five and maybe even three years before; however, a number of Australian States had just gone through a period of very revealing and damaging corruption inquiries. State funds institutions having suffered massive financial loses, private enterprise style undertakings by State government bodies having failed miserably and having cost the tax payers many hundreds of millions of dollars and to top it all off Griener offers a back-hander to an individual to get him out of the way.

Premier Greiner probably doesn't, to this day, believe he did anything wrong and probably given the historic levels of political morality he hadn't. Where he failed was in reading the level of disgust that the public and independent politicians (in the parliment it was independent politicians that forced Premier Griener to resign by having the balance of power and threatening to call a no-confidence vote; which Premier Griener would have lost) held for persons who now displayed these levels of morality. Public morality had evolved and changed, Premier Greiner and his political machine, and in all probability his political advisors, had not; thus spelling disaster.

Lance Chambers remembers from his days as a graduate student of finance the point being made by one of his lecturers that no investment is ever secure forever. The lecturer went on to make mention of a number of cases in history that seemed rather silly at the time of the lecture. One gentleman had tied his legacy, to his children, to the tram-cars of San Francisco, another had invested all his money into the railroad and in his will had insisted that the money must never leave the railroad since his inheritors were not clever enough to make sensible use of HIS money. As you can guess the beneficiaries of these monies struggled for many years to overcome the stupid conditions of these wills, all to no avail and they lived to see significant fortunes slowly dwindle and die as these industries slowly collapsed with the passage of time and the changing needs of the societies they served.

However stupid those men may appear today, when they made the decisions they did it may have really seemed as if the tram-cars and the railroad would live forever. In the not too distant past there would have been those who believed the same for the semi-conductor, computer, automobile, aircraft, etc., etc. industries.

What we are saying is that not only do people have limited corporo-genetic lives but the same holds true of industries. They have limited econo-genetic lives. In fact the nature of nature is that nothing will last forever and the work in GA's highlights this reality.

It may well be that successful planning is invisible? Its success might be reflected by the degree to which a community sees no problems with the operation of its physical infrastructure, services, and the satisfaction of its goals and standards. Given that most socio-economic systems in the modern world are dynamic, evolving over time, developing new facets and discarding old ones, successful planning by the definition of invisibility might be infeasible.

It is certainly difficult to achieve wholehearted success without considerable and continuous attention by those who formulate the plans and those who make the decisions for their implementation. On the other hand, failure in planning is immediately obvious. Infrastructure, services and physical facilities inadequate for the magnitude of the tasks to be performed, or unable to cope with nature of those tasks, provide clear and cold manifestations of an inability to forecast, mitigate or anticipate demands. This is failure in planning. Thus absolute success in planning might just be elusive, but perhaps a reasonable probability of success is achievable. What are the requirements for planning to seek success, or at least avoid failure?

This chapter attempts to describe the nature of the planning task and the role it fulfils in modern society. This role is that of providing information for decision making, noting that those who provide the information (the planners) are seldom those who make the decisions (e.g. the elected representatives at least in most western societies). The chapter then considers the variety of factors that can operate to reduce the effectiveness of planning, or provide impediments to the application of good planning principles and practice. Many of these factors are psychological, based on the outcomes of human beliefs, perceptions and actions, and applying at the levels of the individual, the group, and the community at large. Thus they are difficult to deal with, but strategies to mitigate their influences are essential. Johnson (1983) provided a succinct description of why things can go wrong in planning, especially planning for the provision of public facilities, services and infrastructure:

Decision makers and planners make bad decisions for the following reasons:

(a) Elementary logical rules are violated.
(b) The social systems they are attempting to make decisions about are not well defined.
(c) The hierarchy of authority is confused with the hierarchy of set aggregation. Some decision makers cynically use these confusions for their own purposes.
(d) Social systems are planned in an inappropriate clock time.

Johnson's paper concludes by defining a substantial set of guidelines for effective planning and decision making, and we shall return to consider these later in this chapter.

Next we must consider the nature of the planning task, the ways in which that task might be accomplished, and the purposes for which planning is needed.

5 The Planning Task

Planners are basically involved with generating alternative proposals and assembling supporting information to be presented to the decision makers who have to consider the consequences of those alternative proposals. To this end, the following five propositions enunciated by Boulding (1974) provide a useful starting point:

(1) *The world moves into its future as a result of decisions, not as a result of plans.* Planning is only effective if it provides information of direct relevance and use to those who make decisions. The information to be provided should not merely include that which is desired by the decision makers, it must also provide the range of information necessary for full understanding and comprehension of the short and long term consequences of the alternatives under consideration;

(2) *All decisions involve the evaluation of alternative scenarios of the future and the selection of the most highly valued feasible alternatives.* The decision making process must therefore involve two significant elements. The first is an *agenda* outlining the alternative scenarios and some analysis of the likely implications of present decision choices and future community values, goals and beliefs. The second is a *valuation scheme* that expresses preferences for the characteristics of the likely outcomes of any one decision. In the case of infrastructure planning, the valuation scheme is often strongly related to community values, goals and beliefs;

(3) *The degree of uncertainty associated with the items on the agenda affects the evaluation and decision strategies and the quality of the decisions themselves.* Decisions concerning future actions are based on assumptions, sometimes explicit but often implicit, about the likely consequences of alternative decision choices and the future state of the region where the decision will be implemented.

Thus, the greater the degree of uncertainty associated with these assumptions, the higher the value that should be placed on decisions that leave future options open;

(4) *The products of planning should be designed to increase the probability of making better decisions.* The planning process thus needs to examine a wide range of agendas, the values, goals and objectives underlying the decision, past decisions that are not considered to have been effective, failures of past predictions, and early warning signals of emerging undesirable consequences, and

(5) *The basic result of planning is some form of communication with decision makers. The outcomes of the planning process are but a small part of the information required by decision makers.* The usefulness of the planning information will be increased if real attempts are made to adapt the products of the planning process to more precisely fit the substantive and interpretive requirements of the decision makers. To want to use the planning results, the decision makers will need to be convinced of its authenticity and relevance. This can only be done by presenting the results in forms that are acceptable to and easily understood by the decision makers .

We should note that Boulding's propositions provide no more than a starting point for considerations of the nature of the planning task. As will be seen subsequently, there is much more elaboration that will be needed, especially when we come to consider the reasons for failure in planning, and what might be done to mitigate against such failings.

The primary implication of Boulding's propositions is that strategic planning work should focus on the information needs of decision makers, and recognise the limited capabilities of individuals not familiar with technical analysis to accept and interpret the information that is produced (Meyer and Miller, 1984, pp.7-9). An important assumption in this approach is that the particular decision makers can be identified! This is usually the case in infrastructure planning, but may not always be so. Those who may constitute the ranks of the decision makers are usually those individuals whose task is to allocate available resources amongst a set of competing needs to achieve certain goals. They may thus include elected representatives who set general policies for resource allocation and the appropriation of funding for specific areas, agency managers responsible for operating and maintaining a facility or service, private sector managers whose concerns focus on the most efficient utilisation of their organisation's resources, and community representatives who seek to safeguard certain aspects held to be important by that community.

On the basis that the planning process is predicated on meeting the information needs of the decision makers, it must be stressed that the process should do more than merely provide the information apparently desired by those decision makers. The information needed by them to provide a more complete understanding of the problem and the implications of the various alternatives must also be included. If there is unpalatable information here, then better that the decision maker is made aware of it before too much commitment has been made to a position that may proven untenable. Of course, this is easier said than done in many circumstances. However, a successful and ongoing relationship between the planner and the decision maker cannot persist if an atmosphere of fear and deception has been created. Successful planning and decision making requires a strong degree of acceptance of the respective roles of the planners and the decision makers and the development of mutual trust between the two groups.

A simple definition of the strategic infrastructure planning process can then be attempted, drawn from the above discussion and that provided by Meyer and Miller (1984). The planning task is the process of:

(1) *Understanding* the types of decisions that need to be made in a particular environment at a given point in time;
(2) *Assessing* the opportunities to be available and the limitations to be imposed in the future;
(3) *Identifying* the short and long-term consequences of alternative choices designed to take advantage of these opportunities or respond to the limitations;
(4) *Relating* alternative decisions to the goals, objectives and consequent policies established for the particular realm of the planning problem, and
(5) *Presenting* this information to the decision makers in a readily understandable and useful form.

There are a number of aspects of this description that merit special attention. Firstly, strategic planning is seen as a *process*, i.e. a sequence of ongoing tasks which, as will be discussed later, may have no definable end point. The technical analysis and modelling considered by some (especially those from the numerate disciplines such as engineering and perhaps even economics) to be the planning task, are in reality no more than one of the components of the process. How these heavyweight endeavours fit into planning is described subsequently, in the detailed treatment of the 'Systems Planning Process' presented later in this chapter. Secondly, the planning task involves the consideration of *opportunities* that will arise in the future as well as the limitations to be faced there. One past view of planning was that it should focus on the identification of future problems or deficiencies in a particular system. Clearly this attention is important. But equally there may be scope for changes to build on current or future successes and gain further benefits from them. Such as pro-active approach to planning offers one means for finding diversity amongst the alternative solutions that the planning process is intended to provide. A central theme of this book is that diversity of alternatives is the way to ensure successful planning and the resilient performance of the physical, economic and social systems to which that planning is employed.

Another consideration for the planning task is that of the *time horizon*. There is a need for planners to consider both the long-term and short-term perspectives. Long-term planning is an ongoing activity that reflects the determination of needs and policy direction. It provides the context within which planning decisions can be made in the short-term. It also forces a recognition that some decisions have to be revisited periodically, to cope with changes in technology, community attitudes, and even the sheer scale of the tasks to be undertaken. To be relevant to decision makers, the long-term component of a plan must be flexible and responsive in scale and scope to the types of decisions that will need to be made. On the other hand the short-term component needs to take into account the immediate needs for satisfactory systems performance. The interplay between the short-term decisions and the long-term is of interest: how much do decisions and actions in the short term influence the future outcomes of the system. For instance, how might the staging of construction of a new factory affect the way that the facility will operate when it is complete?

What is emerging from this discussion, and will be considered in more detail subsequently, is that it is this process of planning, rather than the techniques employed in the process, that is of primary importance. Planning is used to provide relevant information for decision making. Poor planning will result when erroneous or irrelevant information has been provided, and this state is indicative of a failure to understand the requirements of the decision to be made and to misread the information and analysis necessary to produce the decision. Planning is not an end in itself, it is the means towards an end.

Conceptual Models of Planning and Decision Making

A theme of this chapter is that planning methodology and the techniques employed in the planning process need to be consistent with the nature and needs of the decision making process if planning and decision making is to be successful. But what is the decision making process? A number of decision making models have been formulated, and these all have their particular areas of application. A brief review of some of these models follows. Following this review, a framework is established for the planning process, as applied to complex systems, that can reflect the needs of a decision making process for use in strategic planning.

Previous research by political scientists and management scientists on decision making processes has revealed a myriad of circumstances and alternative means for decision making (Samson 1988). In all organisations there are different levels of decision making, involving a wide range of participants and many different form of information input. Some of the differences to be considered include:

(a) The type, frequency, structure and complexity of the decisions;
(b) The characteristics, abilities and requirements of the decision makers, and
(c) The institutional and political contexts in which the decisions are needed.

Given these differences there is little chance that any single decision making process will be superior to all others. Rather, any one process may have its particular realm of application, but this may depend on circumstances which vary over time. Five major conceptual models of decision making have been identified (e.g. see Meyer and Miller, 1984), to which most of the identifiable decision making methods may be ascribed. These models are:

(1) The rational actor model;
(2) The satisficing model;
(3) The incrementalist model;
(4) The institutional process model, and
(5) The political bargaining model.

These models come from separate backgrounds, and this must be borne in mind when they are compared. The first three models were developed largely on the basis of one-off decision processes, i.e. the particular decision seen in isolation from all others. The institutional process and political bargaining models were based on the premise that decisions must be taken within the context of the organisational and political settings relating to the environment in which the decisions are needed. Though these models are quite different, they do share some similarities. Further, they are not mutually exclusive, there is considerable potential for overlap between their areas of application.

The importance of the these conceptual models is that they allow planners some insights into the ways by which the outcomes of the decision making process can be interpreted.

Rational Actor Model

The model is a product of the modernist period of the late nineteenth and early twentieth centuries, that saw the application of scientific method as the means to provide logically consistent, socially equitable and economically efficient decisions. It is based on the idea that decision making should seek to maximise the attainment of a pre-determined set of goals and objectives, conducted by a group of fully informed decision makers. This method is the basis of the idea of 'rational decision making'. Its major drawbacks are its assumptions that full information will be available to those making the decision, and that this information will be applied strictly to meet the aim of optimising the level of attainment of the clearly expressed goals and objectives. Can the decision makers really be sure that all of the factors affecting or affected by the decision have been considered, let alone properly described?

The method puts a strong emphasis on the use of quantitative information and the application of models to process this information. Its application is based on a formal operational structure involving a single sequence that identifies all of the feasible alternatives, compares these alternatives against a set of evaluation criteria, and ranks the alternatives in order of preference with respect to the set goals and objectives.

The major criticism of the model is that it implies a certain naivety, failing to recognise the role of human and institutional perceptions, prejudice and perspectives in decision making: its basic premise is that a rational, logically consistent decision framework, free of biases, can be established. Nevertheless, the model almost certainly provides the starting point for its alternatives, and is still seen the idealised method for decision making.

Satisficing Model

The rational actor model is intended to deliver the globally 'best' alternative as the solution of the particular problem. In most circumstances this is unlikely to be realistic. For one thing, there must always be doubt as to whether or not all of the possible alternatives have been properly identified and described. Thus there is no proof that the best alternative amongst the set considered is indeed the global solution. The satisficing model acknowledges this deficiency, and seeks to form its selections on the basis that, to be chosen, an alternative must satisfy some minimum level of acceptability in terms of the evaluation criteria and the goals and objectives, or those which cause the minimum harm or disruption whilst delivering some ascertainable benefits.

So, the basis of the satisficing model is still the rational approach, but with the explicit recognition that there may be limits on the range of alternatives and their characteristics. The decision process is then intended to identify the most satisfactory of the identified alternatives, and to select this as the solution to the problem. The model is concerned with a sequential process that examines alternatives and their consequences, acknowledges that the full implications of each alternative may not be known, seeks to satisfy the goals and objectives to at least a minimum standard of acceptability, and provides a framework capable of repeated application. It is a method of compromise based on rational criteria.

Incrementalist Model

This model attempts to move away from the search for a globally acceptable solution, reasoning instead that decisions are made on the basis of incremental or marginal differences in their consequences - that each decision is but a small part of the big picture. It presents a limited strategic approach, in terms of the number of alternatives considered and the prediction of their consequences. The attainment of the goals and objectives is seen as less important than providing some relief for the problem at hand. In this respect the model is 'remedial' in nature, rather than seeking to maximise the benefits to be achieved. It also assumes that key decision makers are remote from each other, and therefore each decision maker can only have a small influence on the overall system.

This is not a model for long-term planning, rather it is concerned with short term influences and solutions that vary marginally from the status quo.

Institutional Process Model

This model considers that most individual decision makers belong to organisations and that the decisions they make will be set within the environment set by that organisation. Institutional structures, both formal and informal, channels of communication, areas of expertise and influence, and operating procedures will influence the decision making process. Goals and objectives, evaluation criteria and analytical methods will all depend to greater or less extent on the characteristics of the organisation and its relationships with the outside world.

The range of feasible or acceptable alternatives may be determined by the institutional view of the scope and extent of the problem. The implementation of any chosen alternative will depend on the extent to which the implementing agency actually has the responsibility, capability and resources to carry out the implementation.

The usefulness of this model is that it offers a context in which decisions and their fates can be seen. Whilst acknowledging the role and utility of strategic planning, it indicates the constraints that may be faced in the selection of alternatives and the implementation of solutions.

Political Bargaining Model

In this case decision making is seen as a pluralistic activity, involving players from a variety of backgrounds who will often have diverse if not conflicting goals and objectives, values, beliefs and interests. The decision making process must then be capable of seeking common ground between the players, involving trade-offs and bargaining. Without some form of consensus and agreement, no workable decision can result, for none will be accepted if it only represents a narrow, sectional view and other players have the capacity to challenge, reject or undermine it. The implication is that the sharing of power and responsibilities between individuals and organisations is an important part of the decision making process, which will influence both broad outcomes and the quality of the decisions made.

What follows is that the political bargaining model must differ substantially from the rational actor model, for the outcomes are unlikely to be 'optimal' in the definitive sense. The process of trade-off and compromise is unlikely to yield a final state that agrees to any large extent with the best solutions available from the perspective of any one of the interested players. The outcomes will reflect those parts of the problem where the players can agree, and those other aspects where agreement has not been possible will have to wait for future discussion.

Can reasonable decisions then be made using this model? Some will argue that this is not the case: stalemate or inferior solutions may emerge. But the pragmatist might well be happy with the outcomes, recognising that the 'ideal' or 'global' solution as seen by any one of the players will have little chance of coming to fruition, whereas an agreed position amongst the set of players at least has the chance of implementation with a minimal degree of opposition. In a society that seeks democratic outcomes recognising that its constituents come from a variety of backgrounds and possess a full spectrum of beliefs and perspectives, the consensus solution derived from political bargaining may be the best way to provide an effective strategy for problem solving. Stephens (1991) presents some lucid arguments for the use of this model and the reasons why the 'rational-based' models may be of limited applicability in contemporary planning and decision making. She argues that the rational actor model and its descendants, though still highly-valued by planners and engineers because they correspond significantly with the ethos and theoretical and professional backgrounds of those groups, are out-moded, relics of the 'modernist' era whereas the world at large, especially the socio-political elements of the modern world, have moved into the 'post-modern' period in which the ideas of rational analysis and choice are found wanting. Post-modernism attempts to grapple with thc feelings of uncertainty, ephermality, chaos and relativism that permeate much of the thinking in the advanced western societies. It tends to reject expert opinion as being either biased or irrelevant to the real world, which it sees as beyond the control or influence of its inhabitants. It is an uncomfortable state for those, say from the physical sciences and engineering, who lay claim to an ability to understand and explain real world phenomena and then to harness those phenomena for the benefit of humankind. A decision making model aspiring to develop rational alternatives and produce identifiably superior solutions has little claim to support in a social context where its basic tenets are not recognised, if not entirely rejected, by large and influential sections of the society. Perhaps this explains the marginalisation of those groups, such as engineers and scientists, in present-day western society. These groups had their chances earlier in this century, they claimed the insights and abilities to solve the problems of the day, they either failed, or produced new problems of a more serious nature for others (e.g. nuclear power, pollution and environmental degradation, desertification through over-intensive agriculture, social isolation in large cities through land use zoning, traffic congestion on urban roads, structural unemployment through technological change, and how many others besides?). In such circumstances a decision making model that recognises that the interests of a range of players and groups must be taken into account, that no one group will have all of the right answers, and that any solution to a problem must come from a consensus, or 'least adverse impact', procedure, will find substantial support. Such a model is represented by the political bargaining model.

The role of planning in this decision making model will of necessity be much broader than those for the previous four decision making models. The required planning process has to provide as much information as possible on the alternatives being proposed by each of the players. The analysis process to support the process must also be characterised by flexibility, and be able to respond quickly with information on alternatives that arise during the conduct of negotiations. Analysis must also be sensitive to issues that are raised by competing players and interest groups, and deliver the maximum possible amount of information in any evaluation, so that these issues can be clarified.

Considerations of Complex Systems

Modern socio-economic-technological systems, the realm in which most infrastructure planning is needed, are prime examples of complex, dynamic systems. They are characterised by their large size, and complexity of the interaction effects between the elements of the system. Often these elements are subsystems within the larger system, themselves forming a hierarchy of interconnected systems. One particular example is a large city, an urban system comprised of a set of systems within it (e.g. transport, land use, employment, health, water supply and drainage, education, regional economy, population, environment), all of which interact strongly with each other, and also possessing strong links to the outside world and its component subsystems. No one individual, group or organisation can claim to be fully cognisant of the full system and the way that it operates and evolves. No one will have complete knowledge of the full system, although a suitable collection of representatives may be assembled to provide significant insights as to its workings. Even then, the full extent of the system and the interactions between its elements may be difficult to describe with any accuracy. The same holds true for organisations themselves - especially the large multi-nationals.

Making sense of all the information that pertains to such a system is the job of the planner, and is obvious one fraught with difficulties. The traditional approach is to try to simplify the system in some ways, isolating some sections of it from the rest, removing other sections, and trying to simplify the extent and intensity of the interactions. The basic concepts of linearity and superposition from engineering have commonly been employed to this end, although the recent developments in the fields of non-linear analysis and 'chaos theory' have delivered some new alternatives for systems analysis. The work of Atkin (1974, 1981) and his associates (e.g. Johnson (1982)) has provided some new insights into how to describe complex systems, but this work does not simplify the task.

Understanding the System

The first job to be undertaken by the planner is to seek out and tap the available knowledge on system characteristics and performance. A vital component of this job is for the planner to remove the blinkers, to accept that no one individual or group will possess all of the knowledge necessary to describe a complex system. Although a fully comprehensive knowledge base is perhaps not possible, given the above discussion, the planner or planning group will need to extend the available knowledge base as far as possible beyond the scope of that immediately available to them, if for no other reason than to check the limitations that exist on the initial knowledge. The problem will still remain, however, of the possible lack of ability of any one person, planner, decision maker or politician, to make true sense of all of the information that can be assembled to provide an understanding of the environment of interest.

This problem will come to the fore when an attempt is made to examine the complexities of a particular problem. A major issue in the analysis of any complex system will be to unravel the complexities, the conventional approach is to simplify either or both the system itself and the problem to be addressed. This is not unreasonable but may mean that, without due care and often in ignorance, an important issue or facet of the problem will be discarded, which can affect both the understanding of the system and how it operates, and remove the opportunity to consider some alternative solutions. This latter deficiency will mean that the possibility of examining some potentially useful solutions is foregone.

The interactions between the factors and components also pose problems. Again, conventional wisdom is to simplify these interactions, perhaps by paring them down to the 'first-order' effects only. Intuition often plays a part in this process, with planners and others blithely believing that they have a good understanding of a problem, based on common sense: the 'it stands to reason' approach. For example, a belief that 'if we increase speed limits we will increase road deaths. Road deaths are a bad thing, therefore don't increase road speeds'. Many analyses are often limited to this example of first-order planning.[35] For example, the class and standard of the road may also play a role in its safety performance. Freeways, whatever their other failings real or imagined, are demonstrably safer per unit of travel (e.g. numbers of vehicles per unit time, vehicle-kilometres of travel, vehicle-hours of travel) than other kinds of roads - at least an order of magnitude safer- and freeways usually have higher speed limits than other roads. so road speed *per se* may not be the vital factor?

The intuitive solutions offered by the 'it stands to reason' approach may well fail to deliver real solutions to planning problems. This is best summarised by the axiom proposed by Jay Forester, the father of the systems dynamics models that achieved international prominence in the 1970s through the 'Club of Rome' and other environmental

[35] See footnote 30 for an explanation of the order of a problem.

studies. Forester's axiom was that 'intuitive solutions involving changes to complex systems will produce counter-intuitive outcomes' (Forester, 1971).

Planning Tools

There is a significant lack of tools to assist planners in the actual actions required of planning and decision-making. What tools do exist are likely to be limited to applications in certain areas and to particular, usually well-developed, types of problems. A severe lack of tools exists for solving new problems, or determining radical and different solutions to problems (e.g. the concept of the 'ideas machine' discussed by Dickey (1989). There has been little success in developing flexible tools that can be applied in a comprehensive manner, and shaped to fit the details and eccentricities of the particular system of interest. For too long has there been a belief that good planners are born and not made. In some regard we agree. An individual needs to have the capability of abstracting complex structures and viewing a problem much like a set of forces, each acting on others and on themselves. Pulling this way and pushing that, the planner needs to be able to visualise these forces and derive the resultant/s. Such skills can be developed in an individual through education and practice, but the best planners will possess a flair and imagination that is more difficult to inculcate without some innate capability.

Some people have this capacity inherently, others need to develop it over time (this can be an expensive process if too many mistakes are made by the planner in the process of learning). What will help all planners is a tool to assist in making decisions. If a tool that displayed the forces (factors comprising the environment), which other forces they act upon and to what degree (inter-factor effects) and displayed the resultant of all these forces and their effects; then we have a useful tool to assist the planner. No such comprehensive tool currently exists, but we are in the process of constructing and testing one possible contender, as is described in subsequent chapters of this book.

Where might a start be made for such a tool and its application to planning and decision making problems? In the next section of this chapter an attempt is made to use one of the last intellectual developments of the modernist period, that can claim to be applicable to problem solving and decision making in a complex environment where there is uncertainty and relativism. This method is *systems analysis* and one of its applications, as described below, is to develop a systems model of the planning and decision making process itself. This model provides at least a framework in which various of the conceptual models of decision making can be applied.

Systems Approach to Planning and Decision Making

Continual technological change has been the hallmark of the twentieth century. Some recent examples of such change that are particularly relevant for infrastructure planning include:

(1) In transportation, the motor car, the aeroplane, and the new developments in intelligent vehicle-highway systems (IVHS);
(2) In communications, fax machines, satellite communications and mobile telephones, and
(3) In information processing, the digital computer and more recently the personal computer and its attendant software.

With these changes in technological capability come new methodologies for application in other fields of professional and academic endeavour. In civil engineering, an area closely related to infrastructure planning, the primary methodological development was *systems analysis*, which has provided a powerful, unifying methodology for the holistic yet detailed examination of many problems in policy making, planning, design and operation of urban and regional infrastructure. Systems analysis can be applied to the planning and decision making process, at least to offer an explanation of how the process works, even if not to guarantee success in those workings.

What is a System?

In Systems Analysis, a system is defined as a set of interrelated objects and attributes. The universe is one such system, so is an atom. Thus some systems are combinations of other smaller systems, which we term *subsystems*. In the realm of urban planning, for example, we see systems for economic activities, education, health, recreation, water supply and drainage, transportation and buildings amongst many others. Urban planners are thus primarily concerned with the built environment, the populations that dwell in that environment, the needs of that community for facilities and services, and the linkages between those facilities and services.

Community needs include social, economic, technical, cultural and recreational in the broadest sense, and include shelter for family living, transport, education, employment, services, commerce, industry, energy, health and well-being, social interaction, entertainment and recreation.

Large and complex subsystems interact amongst themselves and compete for available resources. A system-wide approach to their planning is therefore desirable, for this provides a useful means to encompass the system in its entirety whilst still allowing analysis of its details. The initial perspective in this discussion is a general overview of the systems approach, which is then followed by a specific systems view of the general problem-solving and decision-making process commonly found in infrastructure planning.

System Decomposition

A convenient way to examine a given system is to decompose it into its subsystems, elements and the interactions between them.

System-subsystem structure is defined by interactions and interfaces between the subsystems. The most efficient structure is one where the number of interactions between identifiable subsystems is minimised. This leads to the consideration of system boundaries. In many planning applications the various components are clearly separated by distinct physical boundaries, but this is not always so. Sometimes the systems analyst will have to make arbitrary decisions on the differentiation between subsystems.

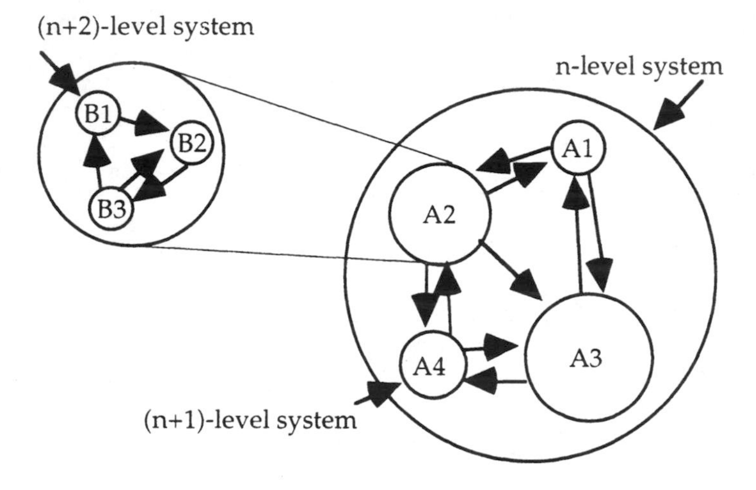

Figure 4.1 A Complex System and its Hierarchy of Component Subsystems

Input-output mechanisms are also important. Few systems are completely self-contained, usually there are some interactions with other systems, with inputs coming from outside into the particular system, and outputs flowing from that system to the surrounding systems. A useful insight into the structure of a complex system is shown by Figure 4.1. This schematic shows a hierarchy of systems and subsystems, which at the n-level may be described as a single element (the system itself as an entity), while at the (n+1)-level the system comprises a number of component subsystems (A_1, A_2, A_3, and A_4) and the interactions between them. At the (n+2)-level the system is further decomposed, with subsystem A_1 shown as consisting of a set of subsystems B_1, B_2 and B_3 and their linkages. The successful application of the systems approach to problem-solving and analysis requires the analyst to develop clear descriptions of the particular system that allow sufficient relevant detail to be included for the purposes of the analysis, without unnecessary cluttering by superfluous detail. Dandy and Warner (1989) provide a useful introduction to the application of systems analysis methods in problem-solving and decision making related to the provision of physical infrastructure.

The Decision Making Process

The Systems Approach provides a powerful method for studying decision making in planning, for it can relate individual decisions on particular projects to the wider view of planning practice and community needs. This permits an investigation of how, where and why some planning decisions can go awry. To start this process of investigation, we need to review the nature of the planning process itself.

As described earlier in this chapter, planners are basically involved with generating alternative proposals and assembling supporting information to be presented to the decision makers who have to consider the consequences of those alternative proposals. Successful strategic planning work should thus focus on the information needs of decision makers, and recognise possible limitations of individual decision makers who may not be fully familiar with the technical analysis that lies behind the planning information that is produced for their consideration. Further, the information needs for decision making may be much broader than just the information apparently desired by the decision makers. Astute decision makers will be equally as aware as the planners of the needs for further information, so that ongoing communications between the two groups can go a long way to ensure that the full information needs can be determined and satisfied. Yet the drawbacks are already apparently. The perceptions, prejudices and horizons of the planners and the decision makers may not be sufficient to recognise just what the full implications of the problem and its apparent alternative solutions might be. The following quotations that have been ascribed to two famous decision makers of recent history may serve to illustrate this common failing of human frailty:

Strategic Planning

> *What, sir, you would make a ship sail against the wind and currents by lighting a bonfire under her decks? I pray you excuse me. I have no time to listen to such nonsense.* [Napoleon Bonaparte, 1803 in discussions with Robert Fulton, the Engineer, concerning steam powered ships]

> *Atomic energy might be as good as our present day explosives, but it is unlikely to produce anything very much more dangerous.* [Winston Churchill, 1939]

Perhaps there are some lessons that are only learnt in hindsight?

The Systems Planning Process

A broad systems representation of the decision making process and its facets was provided by Taylor and Young (1988), pp.2-4. The process, known as the Systems Planning Process (SPP). The SPP is described by the systems diagram shown in Figure 4.2. It may be used as a framework for solving a wide range of problems relating to infrastructure planning, and is equally applicable in the public and private sectors and for problems large and small.

If a starting point can be defined for the SPP - you will see from Figure 4.2 that this is not immediately obvious, nor indeed should it be - then this starting point is the *definition of the problem(s)* under consideration. This is perhaps the most important single component of the planning process. Very often the careful definition of the problem will greatly assist in suggesting possible solutions. Indeed, the simple enunciation of the problem may well be a crucial step in solving the problem itself.

Problem definition may be seen as a part of an internal loop within the SPP, involving the values and goals of the community, the setting of relevant policies to support those values and goals, and the subsequent diagnosis and examination of potential problems facing the community once those policies are established. In addition, the values and goals change continually, so that policy settings and the nature of problems will also need to change.

So, what is a problem? One definition is that a problem is *simply the difference between what one has and what one wants* [De Bono, (1969)]. A more explicit definition of a problem may be given as:

> *a problem for an individual or group of individuals is the difference between the desired state for a given situation at a given time and the actual state. This difference cannot be eliminated immediately (if ever).* [Dickey, (1975)]

There are four significant components in this definition:

(1) *The problem is the concern of a finite population and is not necessarily of general concern to the population at large*. This fact is of some importance when attempting to define the 'goals of the community', for there is no single set of goals, they are different for everybody. This disparity is often a source of considerable misunderstanding since those people not directly affected by a particular problem may find it difficult to comprehend the nature and/or magnitude of that problem and the intensity by which it affects others;

(2) *The identification of the desired state may itself present a problem in that it may be very difficult to determine just what that desired state is, or should be, for a particular situation*. For example, the identification of a desirable level of air pollution may present problems because the desirable state may not be represented by a zero pollution level. There are finite levels of air pollution that are easily tolerated by humans and other living organisms. However, increasing air pollution above some threshold level brings with it substantial problems. The determination of these threshold levels, as a possible representation of a desired state, may involve considerable difficulty;

(3) The observation that *the problem may never be solved* makes many planning problems somewhat different from the problems encountered in other areas (e.g. the physical sciences and engineering)where problem solving skills are highly developed, yet may not be appropriate for direct application in socio-economic systems. This insolubility is due to the complex interweaving of various planning problems and the open nature of the systems planning process. Thus, the best solution for one problem may well create problems in other areas. The conflict between the destruction of tropical rain forests in third world countries and the needs for those countries to find economic developments is one example of this phenomenon. Alternatively, solving a problem for one group or group of individuals may create a problem for another group, e.g. traffic management measures in a local suburban area where, rather than be solved, a traffic problem may merely be moved from one street to another, and

(4) *Problems occur (or are seen to occur) at a given time*. This is because the nature of problems changes over time. The advent of the motor car in the early years of this century was heralded as

the solution to the (then) problems of urban congestion and pollution! Further, even if a problem could be solved, the solution of that problem could create a rise in the level of expectation of the population in question such that a gap would once again appear between the desired and the actual state. This dilemma can be discouraging for the planner and the decision maker since it implies that many planning problems can never be solved 'once and for all', which may fly in the face of populistic politics.

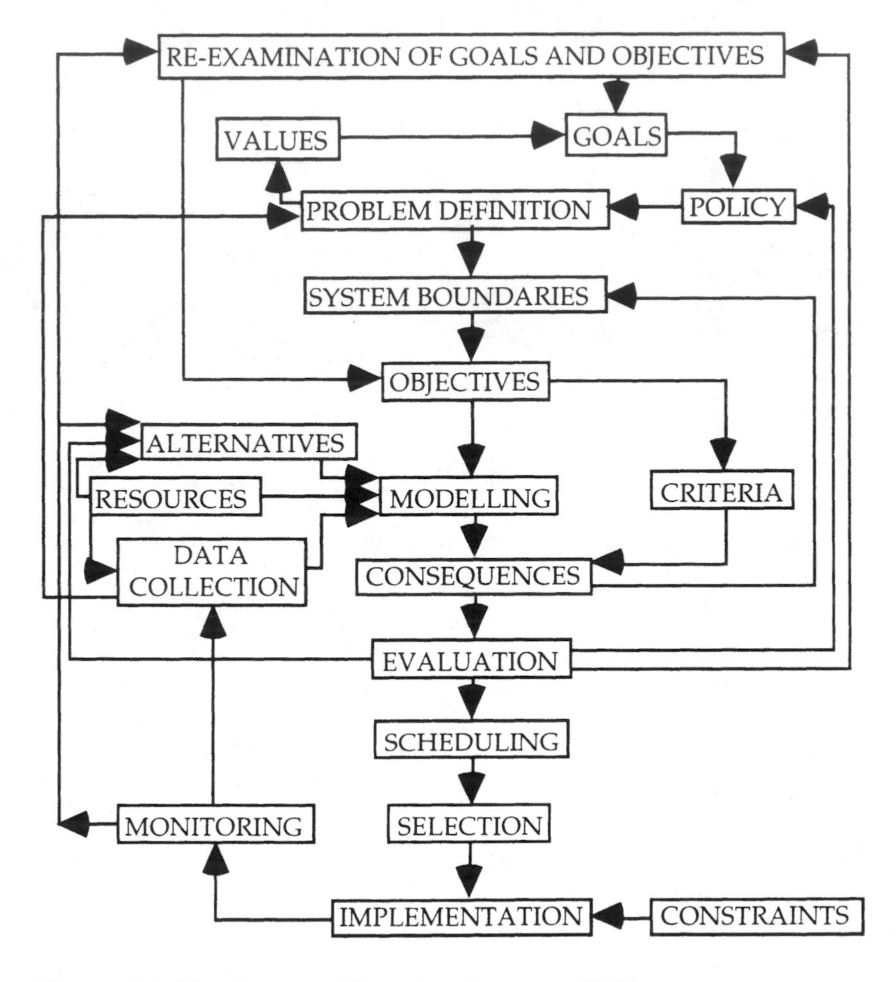

Figure 4.2 The Systems Planning Process (SPP)

Since a problem has been defined as a discrepancy between the desired and actual states of the system, it is obviously necessary to ascertain these two states before attempting to define the problem. The desired state of the system may be reflected in the values and goals of the community. According to Stopher and Meyburg (1975), a useful distinction may be drawn between values, goals, objectives and criteria. Each of these elements represents a restatement, a refinement in more precise terms, of the preceding element. To attempt problem definition, both the values and goals must have been defined. Thus, for example, if one of the values of the community was defined as being 'Environmental Amenity', then one possible goal might be defined as 'the containment of traffic noise levels'.

To determine whether a problem existed with respect to this goal, it would be necessary to ascertain whether, in fact, traffic noise levels were presently considered to be satisfactory. This requires some form of data collection to determine the present state of the system. At this stage of the planning process, the data collection might be relatively informal. For example, have there been any letters of complaint about traffic noise levels from residents of a given area sent to a local municipality or to the state highway agency?

If there is a discrepancy between the desired and actual states then a problem may be said to exist with respect to this aspect of the system. Whether or not a discrepancy exists, other aspects of the system should be examined to determine whether problems exist with respect to the other aspects.

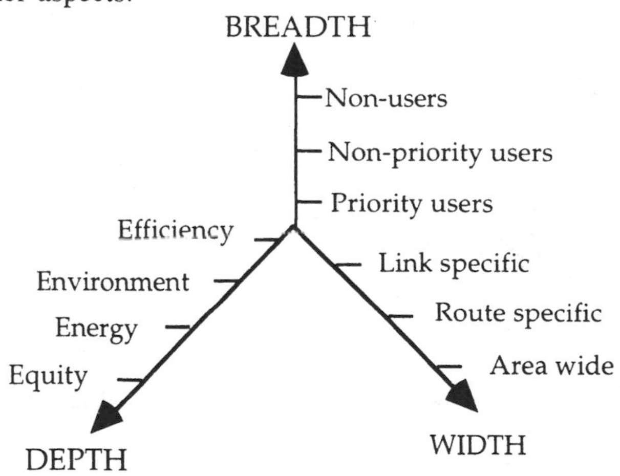

Figure 4.3 An Example of the Dimensions of Evaluation

Having determined the existence of a problem or problems, the next step is to define the *system boundaries*. These boundaries specify the limits of the system to be considered in seeking a solution to the identified problem. The initial definition of system boundaries is made on the basis of an intuitive assessment of the extent of the likely impacts of the problem and may be subject to later review and revision. These boundaries can be considered in an n-dimensional space. For instance, Figure 4.3 shows a three-dimensional space used to specify the geographic boundary of the system (what area does it affect?), the socio-demographic boundary of the system (who does it affect?), and a third boundary that specifies the number of factors to be considered in the analysis (what does it affect?). In this case, the dimensions in this space may be pictured as three axes (WIDTH, BREADTH and DEPTH).

This particular figure was devised for the examination of a problem in public transport planning. The width of an evaluation refers to the geographic areas considered in the analysis. In terms of the evaluation of a public transport service, such areas might include:

(a) Specific links along which the service operates, or the specific stops serviced;
(b) The overall route containing the service, and
(c) The entire area served by the public transport mode.

The breadth of an evaluation refers to the number of groups in the community considered in the analysis. Again with reference to the public transport service problem, such groups might include:

(1) Users of the transport service who have some priority use for it (e.g. transport disadvantaged groups such as the young or the elderly, households without cars, etc);
(2) 'Non-priority' users, such as those who have a readily available alternative form of transport;
(3) Non-users of the system who are directly affected by it, e.g. other road users, or adjacent residents and land owners, and
(4) The general community.

The depth of an evaluation refers to the number of factors that are considered to be affected by the proposal, and which are included in the objective function to be optimised by the implementation of the service. It therefore specifies those factors in which changes are considered as benefits or disbenefits as a result of the public transport service. In very general terms, the factors that might be considered in the evaluation of a public transport service include:

(a) Revenue (fares), cost (operating expenses) and subsidies;
(b) Individual mobility and accessibility to activities such as employment, schools and shopping;
(c) Travel times;
(d) Fuel consumption;
(e) Pollutant emissions;
(f) Safety and personal security;
(g) Equity, and
(h) Capital and maintenance costs.

Further dimensions can also be added, for example *time* may also be an important dimension. Thus we could specify whether an evaluation included effects which occur in the short term, the medium term, or the long term. This definition of the time horizon for a project evaluation can dramatically affect the outcome of the evaluation.

The rigour with which system boundaries are defined will vary with the particular scheme under consideration. In general, the scope of the evaluation should be commensurate with the scale of the project.

Once the system boundaries have been defined it is possible to refine the goals into more specific measures, termed *objectives* and *criteria*, that can be used later as a basis for comparison in the evaluation phase of the planning process. Continuing with the traffic noise example cited previously, the objective might be 'acceptable traffic noise levels on sub-arterial roads in inner urban areas', whilst the final criterion might be 'a 68 dB(A) L_{10} (18 hour) noise standard one metre from the building facade for residential buildings along Major Road in Hushville'. At this stage there is a specific numerical description of the desired state to work towards in seeking a solution. Obviously, other criteria could equally well be specified for this aspect of the problem ...

The *allocation of resources* to the investigation and solution of the problem is a critical step in the SPP since it affects several other key components of the process. this allocation, in terms of financial, resource and personnel commitments, will determine the type, number and scale of feasible *alternative solutions* that can be considered, the amount and type of *data* that can be collected, and the degree of sophistication needed, and possible, in the *models* to be used in the analysis.

The generation of alternatives is perhaps the most challenging part of the SPP from the professional viewpoint. It requires considerable creativity on the part of the planner to generate alternatives that will satisfy the desired criteria within the constraints imposed on the problem solution. The range of alternatives to be considered is wide, and may include one or more of the following:

(a) Do nothing;
(b) Change the technology;
(c) Construct new facilities;
(d) Change the methods of operation;
(e) Change the legislation or regulations affecting the system;
(f) Change the pricing structure, and
(g) Change public attitudes.

It is possible that some professional groups (e.g. engineers and economists) may not the best people to undertake the task of generating distinct alternatives. Whilst generally good at analysis and computation, these groups are often not well-equipped mentally to devise new alternatives. Their minds are too well-ordered to be able to create radically new solutions to old problems. De Bono (1967, and subsequent) would classify them as 'vertical thinkers' rather than 'lateral thinkers'. Given this (possible) limitation, planners with a background in such disciplines should be receptive to, and indeed deliberately seek, the input of others at this stage of the planning process.

The formulation of the systems models to be used in the planning process is governed by the objectives of the analysis as well as the available resources. We may divide models into three basic types: supply models, demand models, and impact models. Supply modelling might generally be seen as a process of determining the effects of changes in the usage of a system on the operating characteristics of that system. Conversely, demand modelling may be seen as the process of estimating the effects of changes in operating characteristics on the usage of the system. Since both of these modelling efforts are concerned with the same pair of variables, with the output of one model being the input to the other, demand and supply modelling form a recursive system, as shown in Figure 4.4. Thus the supply model takes account of the physical characteristics of the system and its current usage in generating a set of system operating characteristics. The demand model then assess the reaction of the users (and potential users) of the system to these operating characteristics, and generates a new forecast of system usage. The modelling process continues until an equilibrium is reached between system operating characteristics and system usage.

Although there is a close relationship between these two modelling efforts, the modelling techniques employed within each of these areas are distinctly different. *Supply modelling* is generally concerned with the modelling of physical relationships between elements of the system. Thus, for example, the determination of system operating characteristics for a bus route would involve consideration of such factors as bus size and performance characteristics, stop spacing, loading and unloading rates, distribution of passenger loadings along the route and general traffic system characteristics.

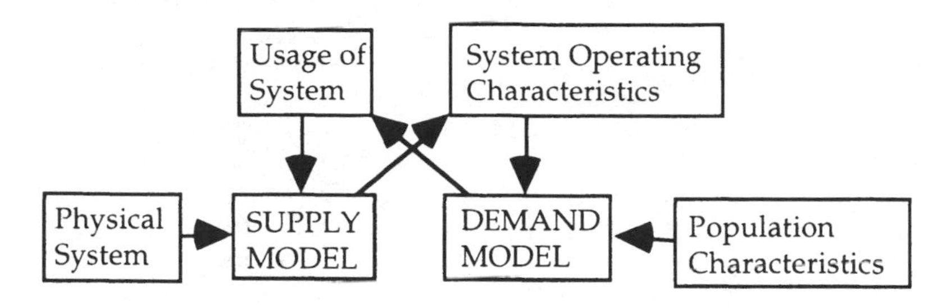

Figure 4.4 Relationship Between Demand and Supply Modelling

Demand modelling, on the other hand, is intimately concerned with the modelling of social behaviour in that it attempts to assess the response of individuals to changes in the physical system. The scope of such system changes and behavioural responses is considerable. Thus, for example, in transport systems planning, demand modelling is concerned not only with the changes in the usage of the transport system itself (as expressed by patronage levels or traffic volumes on various modes), but also with changes in the demand for particular land uses, in terms of trip destinations or residential locations, as a result of changes in the transport or land use systems.

Impact models predict the impact of one system on another, in terms of characteristics that will not affect the demand for use of the first system. For example, traffic noise prediction models estimate the noise resulting from a given traffic volume and traffic composition along a road. It is unlikely that this noise will deter drivers from using that road - more likely the driver's response, if any, will be to wind up the window and turn up the stereo volume! Thus traffic noise models are impact models.

In many cases the 'model' is no more than the experience or guess of the particular analyst. In other situations, the model may be a complex set of mathematical equations or a computer program that can take account of the many system interactions. Whatever its level of sophistication, a model is never more than a means to transform held data (model input) into desired, useful information (the model output). Interpretation of the outputs remains with the modeller and the analyst. In all cases, the model simply makes predictions of the likely *consequences* of the alternatives to be analysed.

The prediction of the consequences of various alternatives may necessitate a revision of the system boundaries if it appears that there are likely to be substantial impacts outside of the existing boundaries. This may also involve a revision of the objectives and criteria used in the process if it appears that goals that were previously satisfied are no longer satisfied. This widening of objectives may then require a change in the models used to predict the expanded set of consequences.

The *evaluation* phase of the SPP simply compares the predicted consequences with the stated criteria by which the alternatives are to be judged. If no alternatives are deemed to be acceptable as a result of the evaluation then a search should be made for new alternatives that fit the stated criteria. If, following an extensive search, there appear to be no acceptable alternatives, then it may be necessary to re-examine the goals and objectives to determine whether they are unattainable and if it may be possible to lower the standards of the criteria without serious consequences.

If one or more alternatives are deemed acceptable then a *selection* of the best alternative is made on the basis of the stated criteria. Selection is also based upon project *scheduling* - what project or parcel of projects can be undertaken in a given time period with a given level of resources. This selected alternative is then *implemented*, provided that certain external *constraints* can be overcome. These constraints include new factors such as political acceptability, institutional rivalry and institutional inertia.

The planning process does not end with the implementation of an alternative. The final phase of the SPP is the *monitoring* of the performance of the implemented alternative. Monitoring is important for three reasons:

(1) It provides data on the actual operation of the alternative. These data on operation and actual consequences may provide the basis for recalibration, or reformulation, of the systems models to enable better predictions to be made of future consequences;
(2) It may also suggest changes that should be made to the selected alternative to improve operation. These changes can then be modelled and evaluated to predict new operating conditions, and
(3) It should be performed to ascertain any changes in values, goals and objectives, and the policies derived from these, which may affect the selection of alternatives over time.

The inclusion of the monitoring step is essential and highlights the fact that planning and professional practice are continual processes that do not finish with the generation of a set of plans for implementation. These plans must be continually revised in accordance with changes in community values and goals, changing economic conditions and developing technology. Successful updating requires not only that planners keep abreast of societal change, it also demands that they have access to significant intellectual and knowledge-based capability that allows for the generation and investigation of a wide range of novel alternatives - for genetic diversity in the search for alternative solutions.

Armed with the products of the above discussion, including an understanding of the planning process and the role of the planner as a participant in the planning process along with a range of other players, most notably the decision maker, there is sense in considering the pitfalls to successful planning.

Pitfalls to Successful Planning

Johnson (1982) asserted that human beings are not very good at making decisions or at planning. He argued that in broad terms, humans are ill-conditioned to consider the essential logic that underlies the decision making process. Primarily this stems from a widespread inability to appreciate that decisions are made in an hierarchical structure, with different groups or individuals making decisions at different levels, and that these decisions are inextricably linked. What seems to be logical at one level in isolation, may be illogical when put into its full context.

Johnson was further concerned about the often inescapable need for speculation in decision making, the requirement to extrapolate beyond the present, observable settings into future settings, and how this affects the quality of decision making. This led him to considerations of time span in planning and decision making, and to a need to distinguish between 'clock time', the passage of hours, days, years in the conventional view of time, and 'social time', the sequence of social events such as births, deaths, marriages, inventions, wars, treaties, revolutions and so on that define the life cycle of an individual or the evolution of a society. Now many of these events can be linked to epochs in clock time, but the general cycles of human history do not relate precisely to clock time durations. Rather, social events are characterised by structure, they are made up of a sequence of many necessary parts that must be combined in the correct manner for the event to have happened. For example, a common event that significantly affects most individuals at least once in their lives is the event of 'moving house'. In Britain, Australia and similar places this event is made up of something like the following sequence of activities:

(1) Put present house on the market for sale;

(2) Inform bank or building society of intentions and desire to negotiate a new mortgage;
(3) Seek details of available properties in chosen area for new home;
(4) Inform lawyer of contractual arrangements;
(5) Negotiate purchase price and settlement date;
(6) Pack up belongings and move.

These parts have to be combined before the event of moving house can be attained. There is some scope to vary the staging of some of the above activities, e.g. (3) might commence before (1) or (2), but this scope is limited. For example, most people seek to synchronise sale price and settlement date on their old home with purchase price and date on the new one. Attaining a subevent such as purchasing a new home without selling the old could lead to severe embarrassment and financial difficulties for most of us. The prospect of selling the old home before a new one has been found might also cause much disruption to a family. Consequently, the settlements usually take place last and on the same day. Some very interesting chains of multiple moving house events involving many households may then be set up! And sometimes these chains fail, when 'something goes wrong' in the event structure, and traumatic consequences result for all concerned!

The importance of social time for strategic planning relates to the evolution of a community's values, beliefs and goals over time. Planning for complex systems in clock time is notoriously unreliable and decisions made on the basis of projected costs and durations can turn out to be quite unsatisfactory. Estimated costs of large systems may well be significantly wrong, especially for government-funded events which do not collapse because of a simple lack of funds. Thus many decisions made for future developments are reliant on the passage of social time events whose precise moment of initiation is unknown, yet must form part of some ordered sequence of events for the decision to be valid. No wonder that many decision makers, especially in politics and business, have a certain reluctance to stray beyond their narrow fields of understanding. Yet without some propensity to speculate, how can future problems be treated by other than remedial measures to shore up a failing system?

For the present treatise, we will assume that the pitfalls for planning largely stem from the following considerations:

(a) A failure to recognise the importance and time-dependent nature of values, goals, objectives and constraints in the planning process;
(b) An inability to generate a large set of alternative solutions possessing sufficient diversity to handle changing environments and belief systems, and

(c) A failure to appreciate the differing emphases and nature that will be ascribed to values, goals and objectives by the different players in the decision making process.

Setting Objectives

One of the major tasks seen in the early stages of the Systems Planning Process is the setting of objectives for a given project. Sometimes the objectives are straightforward, whilst on other occasions they may be quite obscure. In all cases, however, it is most informative and useful for the planning group to state exactly what the underlying objectives are (as they are perceived). This not only helps in guiding efforts towards a problem solution (i.e. attaining those objectives) but also assists in identifying possible conflicts in basic goals between the various players involved in the decision making process.

Goals and objectives may be couched in a broad social, environmental and economic context, and may later be narrowed down to specific technical aspects. It is useful to distinguish here between the requirements of evaluation studies in the private and public sectors. The Systems Planning Process model may be applied in both cases. The major difference lies in the definition of objectives to be used in each type of study.

In the private sector, the system boundaries will be drawn around the company doing the evaluation. Its objective in a capitalist economy is to maximise the wealth of the company, and this will involve maximising the benefits accrued and minimising the costs incurred by that company. In many cases the costs and benefits can be expressed simply in dollar terms. The company is not particularly concerned with who, outside the company, loses when the company wins.

There is much more difficulty in defining system boundaries and objectives for public sector project evaluation. In theory this should be concerned with assessing the costs and benefits to all members of the community. Given this all-embracing definition of a system boundary, public sector planning needs to be concerned not only with monetary costs and benefits but also with various social, economic and environmental impacts of projects.

Whilst public and private sector planning may therefore appear to be substantially different, they are in fact very similar if one thinks of public sector evaluation as being a special case of private sector evaluation, where the 'company' may be a municipality, a state or a nation. The residents of that jurisdiction are equivalent to the shareholders and/or employees of the company, whilst the politicians are a Board of Directors. The Board of Directors is charged with maximising the profits accruing to the jurisdiction whilst keeping the shareholders and employees content!

From the national perspective, economists have often seen public sector objectives as falling within four broad categories:

(a) Maximise national income;
(b) Increase average consumption per capita (i.e. raise the standard of living);
(c) Increase (or decrease) employment, and
(d) Redistribute consumption.

The maximisation of national income (often defined as the Gross National Product (GNP), with an Australian eccentricity to term this income the Gross Domestic Product, GDP) entails an attempt to make the nation perform as a more efficient and productive 'company'. To do this it must be able to increase its overseas exports faster than it increases its imports, since the nation as a whole can only make a 'profit' if it can obtain a favourable balance of payments in international trading activities. Like any company it may be necessary to re-invest a substantial portion of these 'profits' to ensure long-term viability. However, doing this will conflict with the objective (b) above.

If profits are re-invested in the company, there will be less available to give directly to the employees (in the form of pay increases or tax cuts). As it is necessary to keep the workers happy (because they are also voters and therefore shareholders) a delicate balance must be sought between maximising GNP/GDP and raising the standard of living.

Both objectives (a) and (b) may conflict with objective (c), that of increasing or decreasing employment. Both options with respect to the level of employment are included since the favoured course of action will depend on the economic circumstances of the nation. Typically there is an apparent greater political concern with increasing employment opportunities, at least from the public positions adopted by politicians, but in some economic circumstances governments deliberately encourage increases in unemployment, for instance to dampen economic activity that seems to be growing too rapidly. Note that increasing levels of employment may well conflict with increasing the GNP/GDP (where, for example, higher degrees of computerisation and other technological change may be needed for greater productivity) or with increasing the standard of living (where everyone would have to share the total pay packet with a greater number of people).

A concern about unemployment may be seen as one specific example of the fourth objective, (d) above, that of redistributing consumption. As well as redistributing consumption between the employed and the unemployed, one could also be concerned with redistributions between rich and poor, between city and country people, between different regions of a nation, between men and women, between young and old, or between the public and private sectors of the economy. Various governments in different epochs may favour one or other of these redistributions. Obviously, redistribution of consumption will mean that the standard of living will rise for some people but must also fall for others, thus violating objective (b). In addition, obtaining a desired distribution of consumption (or profits) will generally only be achieved at the expense of a lower overall increase in profits, thereby conflicting with objective (a).

Every planning project could be used to fulfil a multitude of possible objectives. Given this, the question remains as to which of these objectives gains priority as a basis for evaluation, and important *who* makes the decision on the selection of objectives and priorities. In this respect, there are three broad possibilities:

(1) Elected decision makers (politicians, company directors);
(2) Appointed decision makers (bureaucrats, planners and engineers amongst others), and
(3) The community at large.

The elected decision makers tend to operate in the medium-term horizon. They are the ones who do the overall specification of objectives and make the final decisions on project implementation. However, it is important that they have a set of objectives that are reasonably in tune with the community (or at least that part of the community that is their electorate). Thus their political objectives (as distinct from their personal objectives) must reflect those of the community if they are to survive into the long-term horizon.

The appointed decision makers are the ones who, in the short-term, specify the detailed objectives of each project. Unlike the elected decision maker, however, there is no need for appointed decision maker to have a set of objectives that reflects those of the community. Indirect pressures, however, may ultimately force recognition of community goals, and thus political pressure can be exerted on planners to adopt certain objectives. Alternatively, the planner may see the need to accept professional objectives (as distinct from personal objectives) which more closely reflect those of the community to be served. Modern planning practice has come to embrace the process of 'public participation' in planning as one means by which planners can determine what these professional and community objectives ought to be. Three major tasks that can be addressed through public participation are:

(a) Information dissemination;
(b) Information gathering, and
(c) Information sharing.

Finding Diverse Alternatives

The discussion of the Systems Planning Process highlighted the importance of generating alternative solutions to a given problem as perhaps the most challenging and creative part of the planning process. The point needs to be reiterated, that the full range of alternatives to be considered is and should be wide. Besides providing more of the same facilities, or repeating previous solutions to similar problems, considerations should be given to changes in the way facilities are used, to changes in community perceptions and attitudes, to changes in other subsystems or systems beyond the direct realm of a particular authority (for instance, there is some evidence that the usage of urban public transport systems can be increased by reducing the general crime rates in a community - not something immediately apparent to or under the influence of transport planners!) Doing nothing is always one alternative that demands attention, if for no other reason than it indicates what the system will be like if procrastination or vacillation persist.

The basic problem with achieving diversity amongst alternatives is the basic inability of human beings to see a wide range of possibilities, beyond their normal scope of vision. Education and professional experiences do not help here either - many professionals lose whatever innate capabilities they may have had for lateral thinking once they are versed in the details of their discipline (whatever that may be) which leaves them expert in a particular field but limited by the established notions, theories, techniques and tools of their discipline. Thus economists will be convinced that problems will be solved by allowing market forces to act, or by setting the right 'price signals'. Engineers will favour technological change and physical construction. Psychologists will look at individual behaviour and perceptions. Politicians will seek to find or respond to expressed community attitudes to seek both possible actions and constraints on those actions. Between the full set of disciplinary inputs may arise a diverse set of alternatives, but this will not happen automatically and explicit consideration is needed. Further, the idiosyncrasies that affect the different groups will need to be taken into account.

Given the difficulties that individual planners and planning groups may find in devising novel alternatives, there is need for other means to generate alternatives. Concepts such as Dickey's 'ideas machine' (Dickey, 1989), intended to allow planners access to a wide range of ideas, postulates and solutions to previous, not necessarily related problems tackled by other people, in other places and at other times, are of interest. Is there ever such a thing as a 'new idea'? As will be demonstrated subsequently in this book, the use of modern information technology and modelling techniques for scenario generation, especially the use of 'genetic algorithms', offers a very real, powerful and accessible means for uncovering and indicating a range of distinct alternatives.

Conclusions

Failures in planning are common and can be ascribed to a variety of circumstances which affect the decision making process. Planning involves speculation about the future state and developments of complex systems, and thus will always involve risks of 'failure', of misinterpretation, limited understanding of underlying processes, and inability to appreciate future changes to social systems. Planning failure cannot be eliminated given these circumstances, but its likelihood of occurrence can be minimised by an understanding of the processes involved in planning and decision making, and by appreciating the very real limitations that humans, even (or especially(?)) human experts, may have in devising and interpreting alternative solutions to planning problems. The adoption of tools and methods that help planners to assemble and examine a wide and diverse set of feasible alternatives, and to look beyond the narrow confines imposed by institutional jurisdiction and the realm of ideas imposed by a particular discipline of human knowledge, are valuable skills whose acquisition should be sought by planners.

6 Why we use Models

The idea - for an **electronic 'scarecrow'** which repels flying-foxes from fruit trees came from seeing a doctor carry out an abdominal scan on the inventors pregnant wife.

- for **cotton buds** came to Leo Gerstenzang when he noticed his wife wrapping cotton on toothpicks to clean their baby's ears

- for **cellophane drinking straws** came to Otto Dietenbach when he was idly twisting the outer wrap from a cigarette packet around a thin steel rod.

Introduction

Man has been using models in every phase of life ever since it was possible to generate the first coherent instinct and models are man's oldest intellectual tool. An autonomous response to a danger signal (such as ducking instinctively) is generated by a model, a mental image used in thinking is a model, a written description is a model. Toys, pictures, films, leggo blocks and words are all different types of models.

Definition of Model

A model of a situation is simply a representation, of our understanding, of the corresponding real world situation, event, process, system or object. Models are idealizations of reality because they are generally far less complicated and sophisticated than the reality they are attempting to emulate and hence are easier to use in helping to understand, for testing, and for research purposes. If variety is defined by the number of different elements in a set, we can now describe the creation of a model as a mechanism for reducing variety from the high number of elements comprising the system being modelled to the much lower number of elements which comprise the model.

The simplicity of models compared with the reality of the situation being modelled lies in the fact that only the relevant factors of the real world are included;[36] that is, all the major and significant features of the situation, suitably modified for inclusion into the model, will be found in the model being constructed. For example in a road map, which is a model of part of the earth's surface, vegetation is not shown because it is not relevant to the use of the map as a navigational aid for motor vehicles. Hence a model is built up of only those factors of the environment that are relevant to the situation being considered at the time. We construct models in this manner so that we can reduce the problem to one that can be managed, understood and to ensure that non-relevant factors do not detract from the immediate task at hand.

The selection process that determines the factors that are to comprise the model will depend on the intention of the model builder who, from a personal understanding of the environment, makes the necessary determinations of the structure and of the properties of the reality to be modelled which will be represented in the final model.

As Haggett and Chorley (1967) said:

> Models can be viewed as selective approximations which, by the elimination of incidental detail, allow some fundamental relevant or interesting aspects of the real world to appear in some generalized form.

The construction of a model makes some statements about an understanding of the workings and structure of the model *vis-à-vis* the reality it is attempting to emulate. Therefore any predictions that come from a use of the model are based upon this understanding which is simply based upon some, now articulated in the construct of the model, theory of the environment modelled.

During the renaissance, Kepler observed the planets in motion and built up a mass of data on their positions at various times of the year. Then, by testing various hypotheses against these data as to the nature of their orbits, he concluded that elliptical orbits nearly fitted the observed data. He thus built up a model which explained planetary motion and, furthermore, allowed predictions to be made about all future positions of the planets. This method of observation and hypothesis has, throughout history, stimulated scientists from different

---

[36] Too often a modelling task can be derailed by discussion and time wasting about irrelevancies. However; it is not always a simple task to determine what is or is not irrelevant.

disciplines to find equivalent predictive models for other particular phenomena.[37]

Functions of Models

Models have many uses, but the most important ones are related understanding and explaining the environment which the model is attempting to duplicate and then is used to make predictions about the future. The forcasting function is probably the most important of all, because, if a model maker has produced a good model of the environment, they would be able to know how the environment would react to different environmental factor changes.

Haggett and Chorley (1967) give an extensive analysis of the function of models:

Psychological function: in enabling some group of phenomena to be visualized and comprehended which could otherwise not be because of their magnitude or complexity.
Acquisitive function: in that the model provides a framework wherein information may be defined, collected and ordered.
Organizational function: with respect to data.
Fertility function: in allowing the maximum amount of information to be squeezed out of the data.
Logical function: by helping to explain how a particular phenomenon comes about.
Normative function: by comparing some phenomenon with a more familiar one.
Systematic function: in which reality is viewed in terms of interlocking systems.
Constructional function: in that they form stepping stones to the building of theories and laws.
Cognitive function: promoting the communication of scientific ideas.

[37] In fact this technique for understanding the world and the systems within which we live did not just occur after Kepler and his work. It is speculated that since the beginning of mankind this has been an on-going process. (continued next page) Ancient witch doctors developed their skills by passing down knowledge gained by observation and by testing these observations against meaningful and understandable, to them, theories of the workings of these systems. In this way whole philosophies, civilizations, and nations came and went. The ancient South American Indians had constructed a fantastically accurate calendar well before this and in ancient England, Stonehenge may have predicted celestial activity thousands of years before.

An outline of the uses of modelling in connection with theory was given by Apostel (1961). These uses include the following:

- The case where no theory is known about a certain group of facts. Instead of studying this group of facts another group of facts about which a theory is well known and which show several important characteristics with the field under investigation is modelled. The model is then used to develop a theory from near-zero hypotheses. Examples of this have occurred in neurology where the central nervous system is modelled via an analog computer programmed to show certain neurological peculiarities.

- The case where a well-developed theory about a group of facts is known but where mathematical solutions prove difficult with present techniques. Using a model the fundamental conceptions of the theory are simulated in a way in which the simplifying assumptions can express the same assignment. Under these simplifying assumptions the equation becomes solvable. In physics the theory of harmonic oscillators in the study of heat conduction is an example of such a procedure.

- The case where two theories are unrelated. One theory can be made the basis of a model whose behaviour simulates the other theory, or else a common model can be introduced interpreting both and thus relating one theory to the other.

- The case where a theory is well-confirmed but incomplete. Here a model can be assigned whereby the theory may be completed through study of the model and its empirical output.

- The case where new information about a certain group of facts becomes available. To ensure that the new and more general theory still conforms to the original group of facts, this original group is constructed as a model of the later theory, and possibly it is shown that all models of this theory are related to the original group of facts, constructed as a model, in a specific way.

- The case where, even if a theory is available about a set of facts, explanation of these facts may be uncertain. In some such cases models have provided explanation (model-use in experimental work on wave theories of light).

- The case where it is impossible to experiment directly with a system because of size, etc. A conceptual model system is constructed and experiments are performed which can be considered sufficiently representative of the original system to yield the required information.

- The case where the theoretical level is far away from the observational level; that is, concepts cannot be immediately interpreted in terms of observation. Models are constructed to constitute a bridge between the theoretical and observational levels.

Types of Models

The following types of models will be described:

Physical	(1) Iconic models
	(2) Analog models
Conceptual	(3) Verbal models
	(4) Symbolic models

According to the how we chose to represent the environment of interest, the model available for use can be grouped into two broad classes: physical and conceptual.

With a physical model, those characteristics that are manifest physically are duplicated in the model. These types of models can be divided into two categories:

Iconic Models

These represent the relevant properties of the real thing by duplicating those properties themselves, the difference is only one of scale.

For example, a drawing of a house can supply a decent duplication of distances and trees, garden, walls, doors, windows, etc., and the only significant difference is the scale of the drawing. Other examples of this class are wind tunnels, tide tanks for testing ship hulls, maps, photographs, and movies.

Analog Models

In these models one property (color) is used to represent another (speed), and hence the necessity of a legend. For example, when we want to show the distribution of speeds on a highway we can use a color histogram that indicates the proportion of drivers in each speed category and supply an appropriate legend which explains the transformation of properties. Slide rules and crash dummies are other types of analog models.

With the conceptual model, the relevant characteristics are represented by concepts (language or symbols). This type of model can be divided into two classes:

Verbal Models

The description of the environment of interest is undertaken using language as the method of communication. These models have limited value when it is necessary to make meaningful predictions or when there is a need to precisely describe or understand the state of a system.

Symbolic Models

In these models, the structure of the environment is represented and expressed symbolically. Models in which the symbols used are mathematical constructs are, as we would expect, called mathematical models. A mathematical model of a system consists of a set of equations whose interactions and problem space explains and predicts changes in the environment being modeled.

Mathematical Models

Components

In the rest of this work, the word 'model' will refer to 'mathematical model' unless otherwise stated.

A model, of the type described, is comprised of: variables, parameters, structural relations, and an algorithm.

Variables

'The activities ... ... which the model tries to represent are called variables. These are quantities which vary over space and time' (Batty 1970).

The purpose of the model and the intentions of the model maker determine which variables are going to be employed in the construction of the model.

Some variables, such as disposable income or GDP, may be connected to economic growth. A variable conceived in general terms (as for example disposable income) must be obtained from an available, verifiable statistic. The variable should be carefully analyzed, to be sure that its role in the model is not compromised or misintcrpreted. As an example, suppose that for family disposable income, we use, in the model, the average of all these collected data for all families in the population. The major problem we can then conceive of occurring is that, if there is a significant spread of these data around the average, that we will not end up with the model performing as we would hope. The model has no understanding of the distribution of family disposable income and only believes that each family has the same level of this variable, as shown in figure 6.1.

In some cases a variable included in the model, because of its theoretical significance, may not be directly observable in the real world, so we must choose a more accessible proxy. An example can be found when we wish to measure 'happiness', but we may only be able to obtain empirically measures such as divorce or suicide rates.

Variables are usually of two types:

- Independent variables (also called exogenous variables): whose values are given as input to the model.
- Dependent variables (endogeneous variables): their values are determined by the model, and given by the model as output.

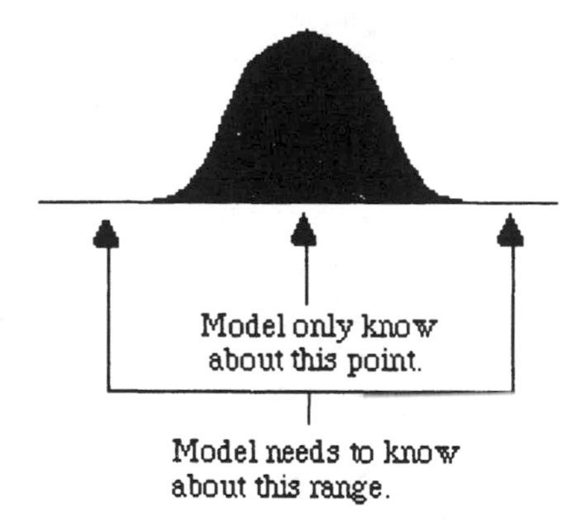

Figure 6.1 Distribution of Family Disposable Income

Parameters

Parameters are values which do not vary within the model; that is, they are constant and unchanging. Parameters are often altered, on a regular basis as new information becomes available, to reflect changes in the environment being modeled. 'They can be regarded as special quantities which adjust the general structure of the model to the particular structure of the world' (Batty 1970). For example, two of the parameters of a population forecasting model are mortality and fertility rates, which are specific to the region under study. As new information on these factors comes to light, from census or surveys, they can be altered to reflect the new realities of the birth and death rates in the region.

Structural Relations

The central core of any model is the particular set of relationships that is developed between the parameters, the dependent and the independent variables, and generally presented as equations. The model makers conceptual and underlying understanding of the environment being modeled is represented by these relationships.

Algorithm

This is the way the mathematics is performed in executing the model or the procedural steps that are undertaken for solving the model. To construct the algorithm means devising a sequence of steps which, when processed or evaluated, will operate on the data in such a way as to solve the problem or produce the desired output.

Diagrammatically, the structure of a model can be represented as shown in figure 6.2.

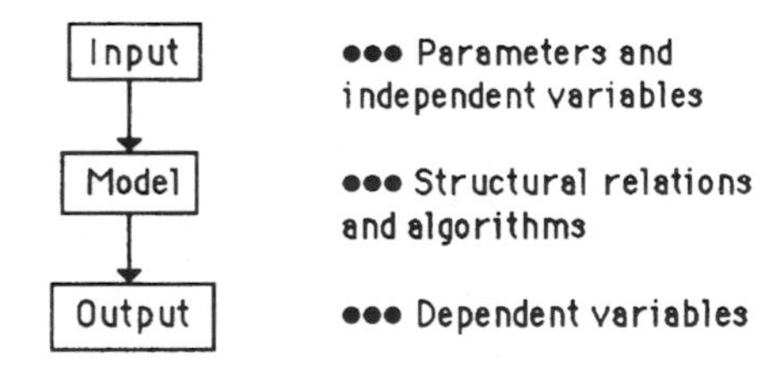

Figure 6.2 Model Structure

Must Models be Complex?

There is one major problem with models and that is that it takes time to construct then. It was pointed out by Clark and Cole (1976) that a primary requirement for a model to be of value is that it should be as simple to construct and understand as possible, the very reason for undertaking modelling, in the first instance, is to allow for the simplification and building of structure of an environment of interest.

Another common problem in the area of modelling is the general complexity of the model construct. This is an abrogation of the true value of modelling which is to develop a construct that can be understood.

The more simple a model construct is, the more easily it can be understood by those who need to understand and employ it. Another great strength is that not only can the recommendations that are derived from the model be better understood by decision-makers, but the process of deriving those answers can be traced through the model and agreement can be easily reached as to the value of those answers. As was pointed out by Senge (1990):

> Simulations with thousands of variables and complex arrays of details can actually distract us from seeing patterns and major interrelationships. In fact, sadly, for most people "systems thinking" means "fighting complexity with complexity," devising increasingly "complex" (we should say "detailed") solutions to increasingly "complex" problems. In fact, this is the antithesis of real systems thinking.

As with many other pursuits, model building, suffers from the law of diminishing returns.

There therefore exists some point at which the effort expended to improve the performance of the model becomes a waste of time and effort.

This problem is most often manifest when there is no strong practical or theoretical foundations for the conclusions reached, when there is insufficient data from which to draw any meaningful conclusions or when there is significant levels of conflict as to the workings of the systems from a set of 'experts' in the particular field.

In these circumstances the development of a 'super model' that answers all questions, is highly complex and unprovable is a patent waste of time, effort, and resources and should be avoided at all costs. However; all too often this is the result when situations of this type are encountered.

Such a model construct adds nothing to our understanding of the environment, of the processes, or of the possible interactions that modify and motivate the 'real world' we are attempting to change.

The attachment of useless and untestable 'features' to models can become dangerous. To foster belief in a model that cannot be shown, let alone proven, to generate any meaningful results is patently a very dangerous undertaking, not only for the decision-makers, those the decisions are made for, but also for the profession of model builders and planners.

Whatever the level of detail, construct complexity and content exhaustiveness of the models, the combined use of high technology and an 'apparently' rigorous mathematical treatment has considerable advantages as an effective tool for propaganda. That the 'success' of the Limits to Growth model (Meadows, Meadows and Randers, 1972) as a propaganda exercise was in a large measure due to the use of a computer has been acknowledged by the project's sponsors, the Club of Rome (Thiemann, 1973).

The bald statement that much of the complexity, that we see in models, is developed for ornament rather than utility is slightly more difficult to justify. It has been demonstrated that there are many models that can be simplified, to a significant degree, while still maintaining the abilities and capabilities required of them in performing the central modelling tasks for which they were designed (Cuypers and Rademaker, 1973). Often modellers will disaggregate their models (that is, add new relationships and variables) so as to anticipate the potential criticism that their models are too simple and are therefore lacking in some respect. This is, of course, a perfectly natural, useful and sensible path to pursue if one wishes to sidestep the potential for disregard for some imagined and apparent failing. But for all that it is too often the case that many of these additions, disaggregations and amendments to the models make no significant difference to the models behaviour and outputs. When this is found to occur it is, more often than not the case, that the modelers will leave such changes in their models. As this situation continues we find that models acquire a great deal of useless and cumbersome complexity. A preferred alternative would be to restore the model to its simpler state after each amendment has been tested and found wanting, while retaining full details of any such amendments which were tested so that one can respond to accusations of trivializing the process of model construction.

Of course, the possibility that a number of 'useless' amendments will, in some way, interact to generate meaningful results should not be overlooked. However, one should never complicate a process unless there is a valid and defensible reason for doing so. Thus we have generated a model that is as simple as it can possibly be; and any accusations of trivialisation can be defended based upon the evidence of well documented tests that were performed during model construction. There are, in fact, very few model builders who have adopted this simplification and testing approach. It is a non-trivial, in time and effort, undertaking and few are willing to spend the time and effort required to construct simple and easy to understand models and modelling systems.

The techniques that currently exist for model construction are not suited to the efficient development of a set of model structures from which to draw meaningful conclusions as to the environment under study. There is value, at this stage, in examining the properties required for a modeling paradigm which would be preferred for the construction of the types of models we have discussed so far. First, the technique should be able to quickly display, in a meaningful and easily articulated manner, the environmental behavioral consequences of a particular model structure. Second, the model system should be able to perform very rapid analyses on the behaviour of the model when based upon different assumptions. These two criteria will enable modelers to construct a series of models by using an iterative process where particular assumptions and constructs can be evaluated rapidly, and accepted or rejected, and by using such tests and evaluations as guidelines for model construction, rather than proceeding by trial and error, as is the current methodology for model development.

The kinds of analyses that, we believe, are required for the construction of truly valuable and useful models includes the identification of dominant feedback processes, stability and control analyses, sensitivity testing, and the full range of available techniques.

These basic requirements, in practice, entail a number of additional needs. The need to rapidly assess the completeness of any model, for example, imposes limits on the complexity of the theoretical relationships which should be included in the model structure. Therefore there exists a need to disaggregate the more complex relational structures into simpler and more easily understood constructs and relationship sets.

Thus it may seem reasonable to allow only linear relationships, since this enables a model to be expressed in an easily understood form. This self imposed constraint is far from being a serious limitation for the four following reasons. First, many of the structural relationships that comprise the models are independent of the particular form of the functional relationships between factors in the models; feedback processes, for example, can be just as easily revealed in a linear model as in any other form. Second, as was demonstrated by Clark and Cole (1976), the overall response characteristics of large complex systems are often quite simple and can adequately be generated by a linear model. Third, many models can be represented in a linear form if a sufficiently small time domain is selected; indeed, Cuypers and Rademaker (1973) pointed out that the behaviour of Forrester's nonlinear WORLD-2 model (J. W. Forrester, 1971) can be quite closely duplicated by an equivalent linear version of a model for a time span of forty years. Finally, we must emphasize that model results and operations are far more sensitive to the factors that comprise the model and interactions between these factors, than to any particular functional constructs of the relationships between these factors. Also, it is worth noting, that one of the most significant aspects of nonlinearity, factor impact time delays, sudden environmental impacts, and phase changes (peace-war-peace, depression-recession-stability-growth, etc.) can easily be incorporated into a linear model construct.

A second requirement implied by our criteria is that the model builder and user should communicate directly with the computer employing an interactive computer modelling system. This would facilitate and speed up the iterative processes involved in model building as outlined above. The present level of model construction technology is such that interactive computing has been applied to enable users, of a fixed and complex model, constructed by 'experts', to experiment with decision-making within this fixed and immutable model rather than to allow the user to construct and modify the structure and relationships that comprise the model itself.

Such an interactive on-line model-building and evaluation system would also need to incorporate facilities for the addition and deletion of model factors and the rapid amendment and recalibration of the factor inter-relationships. One significant 'spin-off' from such a system would be that, providing sufficient attention is paid to the design of the man-machine interface, the potential user of a model could participate directly in the modeling process. This would be in contrast to the usual situation in which potential users of such models are presented with the system structure and relationships as a *fait accompli* and where the user is considered as an 'amateur' and is therefore not to be trusted in the task of model construction and it is not necessary for the user to understand the model or the environment which it is attempting to emulate. It is probably this one factor that has had the greatest impact on the lack of trust to which models and model-builders have been subject. The 'black-box' paradigm is disappearing, and not before time. A decision-maker will no longer trust a model constructed by 'experts' and neither will the public for whom the decisions are being made.

Constructing an interaction matrix, from which a linear model can be reconstructed, requires little technical skill; potential users should suggest components for the model in terms of thier own priorities, objectives, and understandings of the environment, and the visual display should highlight the structural consequences of the inclusion of specified factors and interactions. Furthermore, this type of model-building situation would tend to encourage the development of alternative model structures. It is possible to imagine a scenario in which individuals with divergent viewpoints could use such a modeling framework as a forum for debate in which irreconcilable differences of opinion could be displayed by a range of alternative, yet comparable, models.

7 Genetic Algorithms

We are continually faced with a series of enormous opportunities, brilliantly disguised as insoluble problems.

Introduction

We find the elegance and efficacy of Genetic Algorithms (GAs), in general, very satisfying and exciting. We have come upon a field of research that is relatively new and has, we believe, tremendous possibilities in answering a wide array of problems that up until now were relatively intractable. In this chapter we hope to communicate some of the fascination that we have with these, conceptually and computationally simple yet powerful, algorithms.

A GA, of almost any level of complexity, could in all probability be executed on the slowest of computers at a speed that would still allow for the eliciting of responses at a rate fast enough for most problems. The calculations and manipulations are quantitatively trivial and yet the raw power of these algorithms to solve complex problems never fails to amaze. In this book we are concerned with the forecasting of possible and probable futures and employ GAs to allow us to discover preferable futures and notify us of changes required, to the environment under study, to bring about these preferred futures.

The application of GAs in this book is only one of a wide array of problems that can be solved, as will be exampled in this chapter.

Strategic Models

Strategic models are examples of systems that are the result of the integration of a wide array of subsystems and feedback loops where literally hundreds or thousands of factors may be introduced that can modify their interactions and results. The core problem and also a frequent problem in many other complex systems is that the optimal coefficients of the system's control factors are not known and there is no straightforward algorithm available to discover them. Traditional optimization techniques depend too greatly on a deterministic relationship between the control parameters and the result. These techniques have improved but are unable to optimize the performance of very complex systems, of which policy questions are an example.

Attempts have been made to overcome the effect of this plethora of parameters that have an influence on performance by the simple expedient of recording all the relevant data and developing procedures to handle and interpret these data. These attempts have avoided the main issue as the manipulators of these control procedures are still required to possess a thorough understanding of the system's basic concepts; an understanding which they often do not have, nor have the time to acquire.[38]

When these traditional solution methodologies do work it is generally because the system being studied can be modeled accurately or when the number of control variables is small in number and can therefore be tested for all possible values. For those systems that cannot fit into the two classes described we required different techniques - we require a solution mechanism that will search the problem domain in a highly efficient way without any prior knowledge of the problem and its construct.

The backwards chaining[39] component of the model presented and exampled in this book is an application of the findings of previous studies of how evolution-like mechanisms can direct and improve an automated process without requiring precise knowledge of the environment being studied or for which decisions are being made.

The discussion throughout this chapter is based on the following general understanding: in nature, species are well adapted to their environmental niche in spite of the organisms' great complexity. If we also have an understanding of the very rich array of possibilities available within the DNA structure we can conclude that nature has 'searched' only a miniscule number of the genetic possibilities that are available for examination. Therefore; these highly adapted species, that we see everywhere in nature, were developed by an examination of only a very small fraction of the total number of possibilities available. The set of problem types that we sometimes find difficult to solve are

---

[38] Expert systems are an example of these types of systems. A 'expert' is needed before the system can be constructed. Where such an expert is unavailable we need other mechanisms for gathering an understanding of the environment and for assisting in the decision making process.

[39] The concept of 'backwards chaining' has to do with the idea that we know where we want to go and the model telling us how to get there. The traditional approach has been to 'play' with the model until you hit on the right actions that will tell you how to get there.

far less complex than the problems confronted by nature in the area of natural selection and adaption and therefore if nature has found an efficient[40] and workable method of solving the problem why should we not attempt to mimic it?

'Search' in this chapter means a process of locating a set of particular solutions to a given problem among a finite number of plausible solutions. A solution is said to solve the problem if it satisfies a given objective function. The set of plausible solutions is called the solution space. The object of a search procedure is to minimize the number of objective function evaluations necessary to locate a satisfactory set of solutions that meet the stated objective. A search procedure can be said to be efficient if the total number of solutions evaluated is small in comparison to the size of the search space. The smaller this number, for a given problem space, the more efficient the search procedure is.

Robustness is a measure of how the efficiency of a given search procedure changes when the problem parameters are changed. If the efficiency changes drastically for small changes in parameters then the algorithm is said not to be robust, if however; the changes in efficiency are small for large changes in parameters then the algorithm is considered to be robust. A certain search procedure may be very efficient in solving a specific problem, but if this efficiency decreases drastically for other problems, this search procedure is not a robust procedure.

The search procedures, in common use, can be divided into three main categories: calculus-based, enumerative and heuristic. Calculus-based procedures use either analytical or numerical models of the solution space as the basis for the search. Enumerative procedures search the solution space exhaustively and in a systematic manner and lack any sophisticated mechanisms. The heuristic procedures attempt to improve on the search efficiency of the enumerative methods without incorporating models of the solution space which are often unavailable.

This chapter introduces a relatively new search procedure which joins this group. Before demonstrating this procedure, two common problems in problem space searching will be presented as illustrations. These problems cannot be solved efficiently and robustly by any straightforward and simple procedure available today. Although they

---

[40] It could be said that nature is not efficient, it took millions of years to get homo sapiens to where it is as a species. But consider that, if each generation is approximately thirty years in the making, then it has taken only a few million generations (computational iterations) to get here. That is, in reality, a very efficient computational method and is far superior to a number of the systems currently in place to solve problems far more trivial than the evolution of mankind.

appear to be unrelated to each other, the problems possess common features that can be of value to the search mechanisms available to genetic algorithms - a computation procedure used for searching complex problem spaces based upon very simplified models of genetics, survival-of-the-fittest, and species adaption, as described in the introduction to this chapter. This computation model, its mechanisms and their effect on adaptation and search efficiency, will be described later.

A Collection of Problems

It is helpful to understand which aspects of a particular set of problems make it difficult for traditional search procedures to solve, but are less of a problem for GAs. This section presents the two aspects of search procedures, efficiency and robustness, in two examples.

The Unknown Function

Function optimization is a good example for an examination of the robustness of search procedures, and the frequent trade-offs that need to be made between robustness and efficiency.

The robustness of a search procedure can be demonstrated in the following way: You are given a coin-operated box with an input point and an output display. The box contains an unknown mathematial function, y = f(x). The instructions for the use of the box indicate that when a dollar coin is put into the box, the box takes the value of x which you have supplied, from the input point, calculates its function value, y, and displays this value of y on the output display.

The box does nothing other than evaluate the unknown function, f(x), and no clues are given as to the nature of this function that resides inside the box. Your job to find the number x within the range of 0 to 1, which will generate the maximum value when evaluated by the unknown function f(x), while spending the minimum amount of money. Let us consider the options. One possible search strategy would be to list all possible values of x and to try them one by one.

This procedure is called an exhaustive search because of the effect it has on your funds, it will exhaust them very quickly since there are an infinite number of values in the range 0-1 and you will require an infinite number of dollars to guarantee that you have the correct answer.

Another method might be to simply try different x values at random and record the best one we find as we go along and then stop after a given number of trys. This strategy is called random search.

A different search strategy might be to again pick several x values at random and to select the one that yields the best answer as the target. Then we continue the procedure by investigating around the area of this target. This procedure is based on the belief that this is the area where the best answer will be found. If we find a better target in this narrowed section of the problem space, we adopt it, as the new target and continue to search again in the area surrounding this new target. This is one search strategy and there are many other possible ones, some simple, others quite sophisticated and difficult to implement.

The difficulty involved in the construction of an efficient search procedure to solve an unknown function is further enhanced when we also seek high levels of robustness which is exampled when a different type of box is introduced. In the new box, a door is added so that the function it evaluates can be replaced, at any time, by a different still unknown function and the number of input points can be modified at the same time. We also take the box to two groups. One group is given the instruction to develop new functions that can be inserted into the box and that these can be functions of more than just a single variable x but can rather have any number of variables, while the second group is given the task of developing a search procedure which requires the minimum search resources ($ in this example) over the entire set of functions and input points that could possibly be developed by the first group. In other words, we require the development of an algorithm that on average over an arbitrary set of functions will require the minimum search resources.

The Traveling Salesman Problem

The first example, illustrated above, demonstrates the robustness that is required from a procedure which is intended to be used in searches of highly complex problem spaces. We now offer another example to illustrate the problem of searching large spaces. Large and complex solution spaces are common in a vast array of real world problems. These types of problems also pose serious technical and mathematical problems, as excessively large search spaces require very efficient search procedure regardless of robustness. Let us illustrate the problem by means of the traveling salesman problem, a classic in combinatorial optimization and found in any textbook of Operations Research. A salesperson has the responsibility to visit each of his clients who are distributed geographically over a given area. The solution to the problem is to discover which sequence of visits should be followed by the salesperson to minimize the travel distance traversed to visit each client once and only once.

The salesperson always starts their trip from the same point (home-base), and then develops a sequence of visits which in their opinion requires the minimum travel distance. The number of alternative routes that are possible grows very rapidly as the numbers of clients in the problem grows. We are interested in developing an algorithm that can develop such a tour within a reasonable time.

The next section will present a simple search procedure which we will show has the potential to address these problems as well as other similar ones in problems that have been considered relatively intractable to date.

The two problems which were presented are simply a very small selections from archetypes of many complex problems, problems which are not fully defined, problems which have many alternative solutions, those which are impractical to evaluate systematically, those whose interactions may change in time, and those which may incorporate significant levels of redundancy and structural duplication. All these problems are difficult to solve and present substantial challenges for traditional optimizing techniques.

A Blind Search Strategy

A natural strategy one would tend to use in complex solution spaces, where there is no obvious or straightforward algorithm to search for the optimum solution, is to try a few guesses and to focus on small regions where further search effort seem likely to show promise. The decision as to where to invest search efforts is usually based on local progress of previous guesses. There is value in this procedure if the previous answers told you something you could trust about the problem domain being searched. The type of problem surface that can be searched using these 'hill-climbing' techniques is given in figure 7.1.

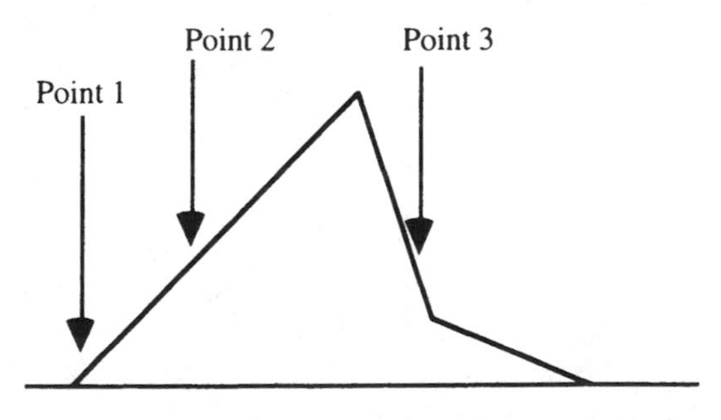

Figure 7.1 Example of Simple Hill-Climbing

Once we have evaluated point 1 and point 2, in figure 7.1 we would search to the 'right' of point 2 since we know that the function increases in value as we move 'right'. After evaluation of point 3 we know that the optimal solution lies somewhere between point 2 and point 3. We therefore concentrate our efforts in searching the space between these two points, in other words we always 'climb hills', in a mathematical sense anyway. However, in complex domains this hill-climbing approach usually proves to be inadequate. Let us therefore consider a different search strategy.

Returning to our original box, to which we now add another box that converts binary numbers to real numbers. The original box is left unaltered. We now simply uses binary representations in place of the real numbers that are fed into the box. The way the binary to decimal converter operates is described by the following function p(g):

$$p(g) = \sum_{i=1}^{l} g_i 2^{-i} \text{ where } g_i = 0 \text{ or } 1$$

For example, the real number 0.40625 can be represented by the binary string 01101, of length 5, because

$$0*2^{-1} + 1*2^{-2} + 1*2^{-3} + 0*2^{-4} + 1*2^{-5} = 0.40625$$

We choose, generated at random, a sample of binary numbers, say 100, and use the modified box to evaluate them (note: from now on we will deal only with the binary representation of the argument x). We link the function value (the real number that is output from the original box) to the corresponding binary number, so that we can associate each binary number with the function value it represents. We are now ready to start the search for strings (from now on we will call the binary number a 'string' to distinguish it from a real number) which correspond to higher function values. The search process then progresses according to the following steps:

Step 1 - Select two members of the sample strings based upon a weighted probability according to their function value.

Choosing a string based upon a weighted probability according to its function value means that on average (and we cannot be more specific that 'on average' because of the stochastic nature of any such selection process) a string that has a function value twice as high as another string shall be selected and duplicated twice as often as the one with the low function value. We then:

Step 2 - Choose at random two positions along the binary string (ie numbers between 1 & l -1 where l = the length of the string) and mark these positions on the two duplicates as cross sites 1 and 2.[41]

Step 3 - Crossover counter-segments between the marked cross sites (Fig. 7.2).

```
  cross site 1|     |cross site 2
parent 1    00|00000|000        0011111000  offspring 1
parent 2    11|11111|111        1100000111  offspring 2
```

Figure 7.2 An Illustration of Two-point Crossover with 10-bit Binary Strings

At the end of step 3 we have created two new binary strings which we shall call offspring (although it is possible that one or both of these offspring can exactly match existing members of the sample, we shall regard them as 'new' for reasons which shall become clear as we continue). We proceed:

Step 4 - Evaluate the function value of the two offspring binary strings through the box.

Step 5 - Choose at random two members of the sample and replace them with the two offspring.[42]

Step 6 - Repeat from step 1.

What we have just described is a very simple model of evolution incorporating natural selection, survival of the fittest and sexual reproduction (summarized in Fig. 7.3).

Before giving an explanation as to why we believe that with certain problems this procedure is better than many other procedures, we shall use our box one more time to solve a problem of maximizing the output value of an unknown function that has been placed inside the box.

---

[41]Note that here we have applied two-point crossover. However; there are a number of differing mechanisms available, see Syswerda (1989), Eshelman, Caruana & Schaffer (1989).

[42] Here we have used random replacement, as for crossover there are a number of different replacement schemes and these are discussed later in the book.

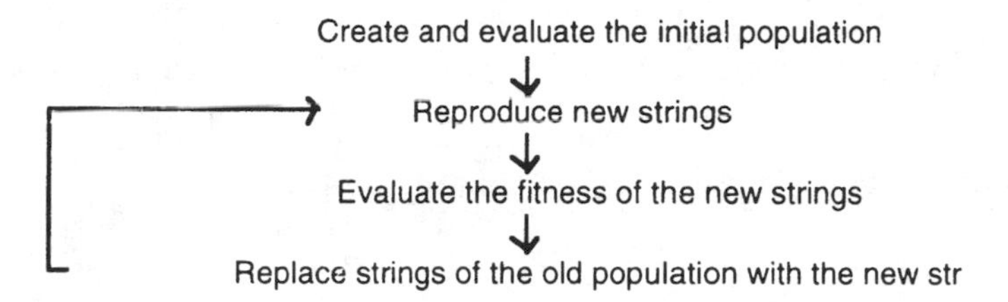

Figure 7.3 An Outline for a Genetic Algorithm

We will start the search procedure by going through each of the steps, one at a time and in sequence, so we can easily follow the processes that are occurring within the GA. Then we will allow the computer to take over so as to speed up the process of convergence towards an answer. We have arbitrarily chosen strings of length 10 and a population size of 8 genes (these values are selected purely for illustrative considerations, and in a real application of GAs they shall generally be much larger. Genes with hundreds or thousands of binary bits and starting populations of hundreds or thousands of genes are possible). A box is given to us with an unknown function inside. We choose a sample population of strings at random (the sample population is summarized in table 7.1). The table shows the eight binary strings (genes) generated randomly, the real numbers these binary strings represent, and the output from the function, hidden within the box, when fed the real number.

Table 7.1 Random Sample of Binary Strings, their Value as Variable x and their Function Value f(x)

No	string	x	f(x)
1	0110100000	0.406250	0.037941
2	1101000011	0.815430	0.069108
3	1101101110	0.857422	0.174605
4	0001100101	0.098633	0.363379
5	1100010001	0.766602	0.000021
6	0110001001	0.383789	0.000974
7	0111010100	0.457031	0.399710
8	1011101011	0.729492	0.000591

The two random numbers, 20 and 67 (generated randomly between 1 and 100), select probabilistically strings 2 and 7. The two cross sites 1 and 6 (picked at random) conclude the reproduction of two new offspring strings (see table 7.2)

The fitness of the two offspring strings are evaluated and they then replace two strings of the sample which are selected at random (table 7.3).

Table 7.2 Parent Strings, Cross Sites and Offspring

parent 1	1|10100|0011	1111010011	offspring 1 = 9*
parent 2	0|11101|0100	0101000100	offspring 2 = 10*

It is clear that such a small sample size is very susceptible to sampling error, so we increased the population size to 100, and let the computer complete the procedure by employing the box to process the strings. Fig 7.4 plots part of the initial sample population, Fig. 7.5 presents the population after 30 generations and Fig. 7.6 shows the population after 100.

Table 7.3 The Sample after Replacing Two Strings

No.	string	x	f(x)
1	0110100000	0.406250	0.037941
9*	1111010011	0.956055	0.000000
3	1101101110	0.857422	0.174605
4	0001100101	0.098633	0.363379
5	1100010001	0.766602	0.000021
6	0110001001	0.383789	0.000974
10*	0101000100	0.316406	0.041796
8	1011101011	0.729492	0.000591

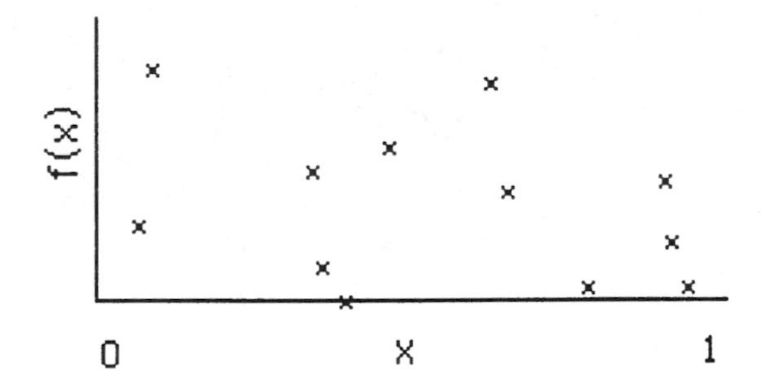

Figure 7.4 The Initial Sample of Search Points Coded with 10-bit Binary Strings

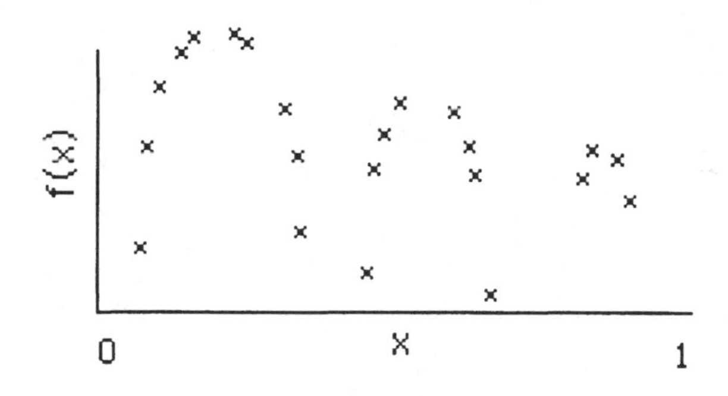

Figure 7.5 The Sample after 40 Iterations

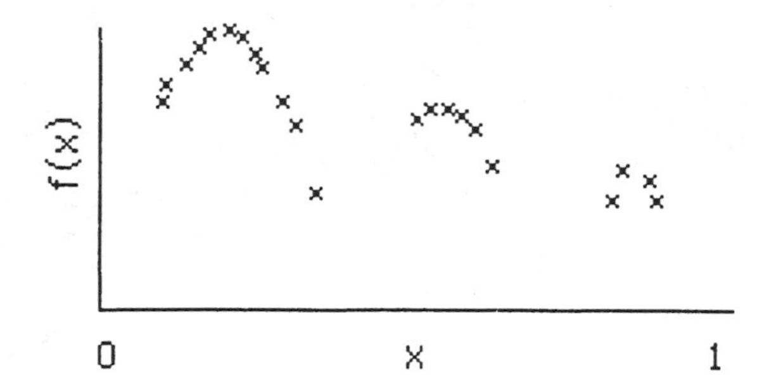

Figure 7.6 The Sample after 100 Iterations

We stop the computations, after we have processed the 100 generation, and we then open the box to check the function which was hidden inside. Fig. 7.7 is a plot of this function (adapted from Goldberg, 1989 pg. 186). Now that we know how the function behaves, we can appreciate the adaption that has occurred and the changes that have occurred from the initial starting population. The capabilities of this genetic like adaptive search process has generated, efficiently and effectively, a very good answer.

There is a clear movement of 'x' upwards and to the left, in search of the 'better' answers. It is as if the GA gained an understanding of the environment within which it was operating (came to know the form of the hidden function in this particular example) and actively generated new genes of fitnesses that were greater than that of their parents. As the process continues the population that survives, on average, becomes better suited to the environmental niche it inhabits. This is what happens in nature and what has happened here in the microcosm of the computer, the genes that we created and caused to be manipulated, the GA and the problem presented to it.

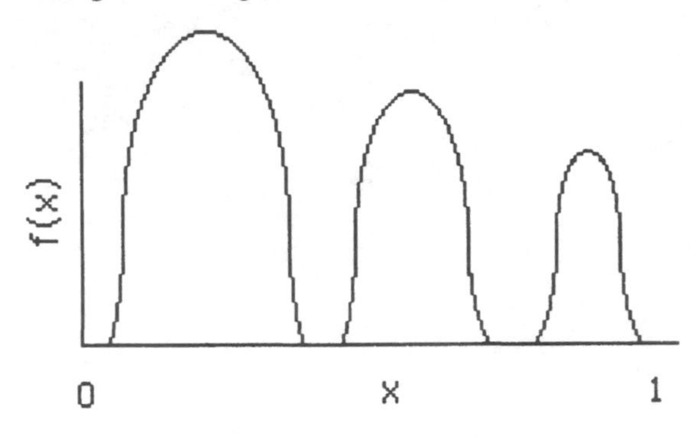

Figure 7.7 The Function f(x) that was Inside the Black Box

Because we never know in advance how good our best guesses are (how fit are the genes we first create), we have need of an algorithm that adapts itself, modifies our genes, confirms its findings, checks the continual improving fitness of the populations it replaces our best guesses with and this is what GAs are and what they do and they are these to the Nth degree and they do these things exceptionally well.

Summary

In this simple introduction to GAs we have attempted to achieve two objectives: to develop a general understanding for the types of problems which it is sensible to attempt to solve using a GA solution approach, and to show how, at the most simple level, a GA operates.

On the basis of the understanding we now have with the mechanisms of GAs we can now also explain and elucidate upon some of the differences between GAs and the other classical search and optimization procedures. These differences are:

- GAs process a coded representation of the problem space rather than the parameter values themselves
- GAs require and generate a population of solutions instead of a single solution
- GAs synthesize information (understanding?) from points that are bred from within the entire space
- GAs operational mechanisms are totally independent of the structure of the problem domain
- GAs do not require problem-specific information or knowledge

One comment needs to be made regarding the illustrated operation of GAs - The number of offspring generated at each generation may not necessarily be 2 as shown, and usually will be equal to the population size or some computed percentage of the population; termed the crossover probability.[43]

Using Evolutionary Mechanisms as Adaptive Search Procedures[44]

Evolution, as the word is used and interpreted by physical scientists implies a process of change with time, but, for biologists, it is more sharply and differentially defined in meaning: Evolution refers specifically to Darwinian selection, a process whereby the offspring of an organism manifest some variation in design. This variation causes them to be selected in preference to others of their species, by being capable of interacting more favourably with their environment, and to increase their probability of survival and reproduction and hence give rise to new varieties that are even better suited to survival and reproduction, in a continual process of species improvement within the environmental niches they inhabit.

Darwins theory of evolution has a problem with the particular behavioral characteristic that is evidenced by the willingness of some organisms to risk and possibly lose their own lives in an effort to protect their offspring. Because, if natural selection works towards the elimination of factors that reduce survival traits within the organism, it should have eliminated the traits that cause the organism to sacrifice itself in order to benefit others. To overcome this dilemma, science brought about a shift in understanding and made the determination that natural selection works, not at the organism, but rather at the chromosome level. This implies that any organism that risks its own survival to improve the chances of its young surviving does serve the case of the survival of the fittest and agrees with the theory of evolution, but this service is given to the organisms genes (as changed and possibly mutated [improved?] in their offspring) rather than as an offering to its personal (older [and less fit?] genes) survival, because the young who carry the parents genes have a better chance to

43 If we have a crossover probability of 80% then for each generation 80% of the existing population will be selected, using some predetermined selection procedure, not necessarily random, to breed. The remaining 20% will not be bred this generation but may be selected, if they survive, in the next generation.

44 For a very good explanation of the use of GA's in the search for a passage in Shakespeare and for the generation and evolution of 'biomorphs' see chapter 3 of Dawkins (1986). This is also a very interesting treatise on genetics and popularises Darwin's theories of evolution.

reproduce in the future and so ensure the survival of the chromosome.[45]

By thus transferring the level at which natural selection operates, it became the genes that were now considered as the basic driving forces behind natural selection, placing the organism with its impulses, desires, instincts, and needs, in subordination to the genetic material it carried and was able to pass on to its offspring – its genes. From this point of view, the 'will to live' is not a defense mechanism inherited to protect the organism, but is rather, a survival mechanism, for the genes, and a mechanism that improves the chances that these genes will be replicated – even at the price of the life of the genes' carrier, as long as the carrier has reproduced. This also means that the 'human' qualities, we are all so proud of, such as love, hate, and compassion only exist because they have survived the harsh tests demanded by natural selection process, and have survived the passage of millennia, because they served the genes', rather than the organisms', interest. Because of these facts we cannot claim that our social behavioural patterns are formed solely by 'culture' and 'civilization' (especially due to their short existence [only 10,000 years or so]), and the 'higher' workings of the human mind. It has to be understood in terms of a process that has been continuing for several million years – genetic evolution.

According to Darwin, however, the basis of evolution is the occurrence of random hereditary modifications in individual organisms of a particular species. The advantageous modifications are promulgated and the disadvantageous ones die out through the process of natural selection. The organism carrying a negative change, a change that results in some survival disadvantage[46] relative to its fellow organisms, will find the struggle for food and a mate more

---

[45] These messages are, of necessity, simple ones for those few organism that are actually capable of retaining these instinctive messages and passing them down, to their offspring, within their genes. It is far more sensible, for the survival of the chromosome, if the message to protect the young only kicked in when there was a chance for the young to survive without the protection and attendance of the parent. It makes no sense for the parent to die protecting their young if the young will simply starve to death once the parent is gone.

The message must be simple. That is why parents will, in some species, die to save day old offspring – a pointless task. The message must be of the form, 'SAVE YOUR OFFSPRING'. A message any more complex, with certain triggers such as: 'Save your offspring only if they are mature enough to survive without your assistance', could not be passed on, from generation to generation, without an increased danger of getting garbled and causing dysfunctional behaviours that might endanger the survival of the chromosome itself.

[46] Note that we are discussing gene survival advantage and disadvantage. Not survival of the organism itself.

difficult than its fellow organisms, and become more vulnerable to the environment within which it finds itself and within which it is necessary to survive and possibly prosper. The Darwinian theory of evolution, with its description of the random nature of evolution, has given us the best rational explanation of the adaptation of species to the many and varied environmental niches that are found in nature.

Differences in offspring reflects differences in the genes, the blueprint from which they are constructed. Interaction of these offspring with their environment and changes in the environment they inhabit determines the survivability of mutated and non-mutated organisms and thus the nature of the later physical manifestation of these genes, the upcoming generations of organisms. Mutations can be brought about by a number of different causes, random gene duplication errors, chemical influences, and radiation. We generally believe that differences in gene structures are essentially random, other than the partially deterministic gene structures brough about by the combination of two genes in a breeding cycle. The selection mechanisms used in nature discards those organisms that do not work out as well as their parents or as well as favorable mutants. Therefore, the mutants and parents that survive and reproduce most successfully come to dominate populations in their given environmental niche.

We now turn to concentrate on the gene itself and the mechanisms it has available for creating the diversity we see around us in nature. It is a fact that genes could have been structured in a number of different ways to allow them to create the degree of diversity possible, of which only a very small part has ever been tried and tested in the harsh arena of evolution and survival of the fittest. There is no gain or loss to the diversity made possible if nature had employed any one of a number of different ways of constructing the gene itself. We will demonstrate this fact by employing a very simple example. Consider the set of integer numbers {0,1,2,...,63}. An octal number (to the base 8) representation requires only 2 characters to represent the full range (8*8 = 64), while a binary number (to the base 2), will need 6 characters (2*2*2*2*2*2 = 64). Independent of how we choose to represent (model) the problem space, using eight's or two's, the full space must be able to be represented by some combination of the way we choose to represent this problem space. Independent of the preceding, the following facts must also be stressed. It is not the solution space (the organism) one attempts to manipulate by representing it in different ways. Rather, it is knowledge about the structure of the solution space (the diversity of organisms that are possible) and the possibility of capturing such knowledge in order to improve the solution (how to create a better organism).

The selection of a preferred representation type depends upon the representations 'capability' of improving our ability to identify and the clarity with which we can identify the manifestation of the problem space. From the preceding discussion we are made aware of the fact that we can select any representation mechanism as long as it is capable of searching the complete problem space. We will lose nothing, independent of the representation mechanism we select as preferable for the problem under study or the methods available for mechanically or intellectually solving the problem, by choosing a tractable representation.

The diversity available for creation does not change with different representations. What does change, when we represent one space by employing some other replicating space, is the perspectives from which we can explore the problem space and the degree of resolution down to which we are able to inspect it. Consider again the binary and octal representations. If we know that a given binary string has a '0' as its leftmost digit we can state that the representation has a value that is at most 31 ($0*2^5 + 1*2^4 + 1*2^3 + 1*2^2 + 1*2^1 + 1*2^0 = 31$). A similar comment can be made if a string has a '1' in the leftmost digit will have a value of at least 32 ($1*2^5 + 0*2^4 + 0*2^3 + 0*2^2 + 0*2^1 + 0*2^0 = 32$), and these comments continue for all the other digits used by this particular binary representation. We can therefore make six such comments for the binary representation since there are six binary values required to describe the complete problem space. However, we can make only two such comments if we employ an octal representation because in this case the representation contains only two digits.

If we wish to comment upon the levels of similarity, or difference, between two or more strings, we can make more comments and gain more information and understanding about the problem space, when the representation is detailed. We can therefore conclude that, *ceteras parabus* (everything else being equal), it is preferable to represent a problem space using a representation mechanism that is as detailed as it is possible to construct. The binary representation is the most detailed representation we can possibly devise for representing real numbers (it should be remembered that we use binary and real numbers throughout the discussion for purely illustrative purposes and it is by no means intended to suggest that GAs are restricted to processing real numbers using a binary representation). Our aim was to acquire insights into the representation mechanisms employed in GAs which closely duplicate the mechanisms employed in nature. Therefore, the binary string representations, employed in the GA to be described in this book, can be likened to the information structure employed by chromosomes and like those natural chromosomes, they have the same capabilities to improve their offspring's' fitness to survive in the environmental niches within which they find themselves.

By accepting a chromosome representation structure, as described above, and manipulating a population of binary strings for a solution search in multidimensional, nonlinear problem spaces, the search emphasis is shifted from complete strings to the interpretation of partial strings,[47] or better still, to the process by which partial strings are identified and differentiated. Indeed, if a better understanding of the problem domain is desired, then it is useful to study the string similarities together with their corresponding fitnesses, and especially to investigate structural similarities of exceptionally fit strings.

> ...In some sense we are no longer interested in strings as strings alone. Since important similarities among highly fit strings can help guide a search, we question how one string can be similar to its fellow strings. Specifically we ask, in what ways is a string a representative of the other string classes with similarities at certain string positions? (Goldberg, 1989, p.19).

Why are they Called Genetic Algorithms?

It is clear from the wide selection of operators available to users of GAs that there is no definitive description of what constitutes a GA, at least not a description that is based on its operators. Furthermore, although the diversity of mechanisms employed in biological genetics is very wide, we have no difficulties in recognizing them as 'genetic' in nature. This suggests that biological genetic mechanisms have some general underlying concepts which are common in many of the mechanisms employed and are easily recognizable as such.

By understanding the fundamental theorem of GAs we realize that it is a procedure which is being emphasized by GAs, not a particular computer coded representation or program. The most fundamental aspect is the specific structure of information. Nature stores information using very simple representations and therefore requires very large numbers of such representations to describe the complex

---

[47] In the search for a single number in the range 0-63 we were interested in the representation given by the whole string. However, in the case where we are searching a multidimensional problem space we may find that variable coefficient values are represented by only portions of a string. As in chromosomes we find different genes for different characteristics. A gene for eye colour, another for nose shape, another for the size of the heart, etc. We also duplicate this representation in GAs. Having single strings (chromosomes) that represent a number of variable parameters (genes).

processes that completely encompass the definition and exhaustively describe the biological entities we see in nature. GAs adopt the benefits that result from these biological information structures, and it is that approach to information representation which allows us to call GAs 'genetic'. Without the structure of information used by GAs and particular to biology, most of the GA operators would lose their meaning and effect.

There is also the idea that the fate of individual solutions carries little importance both in nature and for GAs is reinforced by the processing that occurs in GAs as they execute upon the information (strings) presented to them. Whether or not the individual solutions are appropriate, if valid sub-solution sets (part of a string) that go towards defining a particular string, of high fitness, have ample presence in the population, then such a sub-string will emerge, as dominant, at some stage in the computation process. If however, any such sub-solution sets have low fitness associated with their existence then the same processes will see these sub-solutions being discarded and bred out of the population of strings employed in searching the solution space.

Lastly, we will concentrate on the role of string recombination by using string crossover operators. It is clear that selection can operate only if it has an array of strings from which to make a selection. The question to be asked in this context is: "How can we create variety?" We realize that nature has species specific gene pools from which to develop variety. This variety is created by the reproduction of species, in this exercise genes are combined by the 'sharing' of gene material to generate a new gene–a new variety. Thus gene material is available for combinatorial rearrangement through reproduction within a population. Thus nature, and a species, has available to itself an enormous genetic reservoir. These facts further substantiate the benefits of a chromosome information structure and the use of recombination as a viable means of creating variety. These natural operators are also realized by GAs and are the main process by which variety is be created in their workings.

Mutations however have a diminished importance in providing variety because of the mutual dependence among genes and because this mutual dependence can result in ratios of the general order of 1:1000 working against successful mutations (ie on average, for each mutation which results in an improved fitness, there are about 1000 mutations which reduce the fitness of the organism). Mutations are needed for operators of secondary importance and in situations where the population size is small and thus prone to large statistical noise. Mutation is also of importance when the strings have 'converged' upon a solution and there still exists a need to explore sections of the problem space. Mutations are a mechanisms that allows for the 'casting-of-a-net' out into the problem space to seek out new viable solutions. To boldly go where no solution has gone before.

When to Use GAs?

When people are new to GAs, most often the first question they ask is: "When should I use a GA instead of some other mechanism?" However, the question is not simple to answer because it depends on a wide array of factors, which are usually not known when the question is in the process of being formulated. We will still, however, attempt to give some directions as to when it is preferred to use the GA approach to solve a particular problem. To do so we have to revisit and reiterate the basics of the genetic approach and present a number of arguments. Although there are a wide array of practical applications and therefore evidence as to when GAs have proved efficacious the theory itself, within which GAs were developed, cannot offer any conclusive answers to the question.

Part of the answer is that each string evaluation in the execution of the GA, in the search for a solution, provides information on the structure of the string itself and the point within the solution space it represents. Another part of the answer is that the larger the problem space being searched, the greater the advantage the genetic approach has over other procedures. The GA technique suggests a search efficiency in the order of $O(n^3)$.[48] This means that for a population size n there are on average n^3 'answers' which are being processed efficiently.

Another important feature of the genetic search procedure is the transmission of global information from string to string (i.e. the partial string results that are being processed and that contribute significantly to fitness of the whole string and that then emerge to dominate the complete population of strings). It is this processing of global information that is so lacking in other search procedures, and it is this global information that provides the dramatic search robustness that is associated with GAs. The global information exchange through GAs suggests that multi-modal, multi-dimensional spaces can be searched and processed with reduced risk of prematurely settling on a local optimal region. In fact, it is only worthwhile to apply GAs to spaces that are complex and nonlinear enough so that the use of the array of targeted algorithms and techniques are unsatisfactory. The search robustness GAs inherently exhibit suggests that when entrusted with arbitrary problems such as different functions to optimize, GAs have, on average, a good chance of out-performing other techniques.

---

[48] For a detailed explanation on genetic search efficiency see Goldberg, 1989.

The fact that ultimately any deterministic problem can be reduced to a list of fitness values, makes the underlying string representation a methodology which causes different real problems to be presented to a GA each in much the same way. Hence, we have a situation where the GA needs not be tailored to any particular problem, by coding the problem as a genetic search procedure each problem is transformed to a another type of problem where each such transform has elements of similarity to every other transform. We are then not operating on the original problem but a transformation that is similar in form to all other transformations. We can therefore use a single process, optimized to solve the transforms, to solve all other problems independent of the differences between them in their original form. Hence the strength is in the transformation mechanism and this can be likened to transforming all foods into water, bringing the water to the boil, and then transforming the water back into the original food but now the food is beautifully cooked and prepared. Therefore the same pan, utensils and heat source can cook any and all foods equally well. This is an analogy of the processes used by GAs to transform, solve and reconstitute problems of many types.

Summary

Now it is easier to understand the motivation in employing a represention of a problem as a transformed space. As a matter of fact, we can increase the efficiency and processing power of the search when we select a representation which is as detailed as possible, as long as the representation is still natural to the problem space being investigated. We have also learned that although the algorithm only manipulates individual and complete strings, it is in fact the 'background' growth and survival of partial strings that determines where future computational effort should, and will, be employed.

Because partial strings (genes) are 'hidden' among a large number of complete strings (chromosomes), they are much less vulnerable to noise and stochastic errors. That is the reason why we can allow for the possible disappearance of very good (highly fit) strings through the GA process of random selection, for replacement, of parent strings in the parent population by newly bred strings that may exhibit low fitness values (eg. replace good strings with bad strings). If the parent strings, that have been replaced in the parent population, are in fact good, then the partial strings they have caused to be reproduced in their off-spring which are left in the population, shall in time, bring about the production of that good string again.

It is very important that this aspect of partial string[49] processing is recognized and appreciated, because it is this aspect that gives the GAs their breadth and depth. It shows up the fact that it is not the individual string that is of concern, independent of how fit it is, but rather it is the gene (partial string) pool from which new populations will be bred. The 'good' gene material will be retained and passed on through the generations to reappear, in new combinations with other genes, to create even better and fitter phenotypes (organisms).

An important aspect of partial strings is that when one contemplates a new application for GAs ('new' means that either the problem domain or the representation of the problem domain is new), the first and primary investigation should consider whether the proposed representation of the problem domain is suitable for a GA approach to solving the problem. In addition to the obvious strengths of partial strings on the nature of the building blocks (that they should be short and of low order), the epistasis analysis gives some notion of the amount of nonlinearity a suitable representation should contain. With the two perspectives on analyzing GA efficiency that have been described in this chapter, we have developed an understanding of the types of representations, and problems, that are necessary prerequisites for an efficient search in a GA computational environment. These types are generally complex and large problem spaces that have some underlying co-adapting relationships in their structure.

If we observe the levels of complexity we find in nature, and the success with which genetics directs the evolution of species to be better fitted to their environmental niche, we can draw analogies with many search and optimizing problems found in other disciplines. For most complex, real-world applications, the search for co-adapted control parameters can be a very tedious, demanding, and often-times pointless, pursuit. The analogy with natural adaptive mechanisms is the backbone of GAs, we strive to determine where the values of a number of co-affecting factors (organism attributes), in a problem domain (environment), are to be set, so as to bring about some desired end (fitness).

[49] In the GA literature these partial strings are termed schemata.

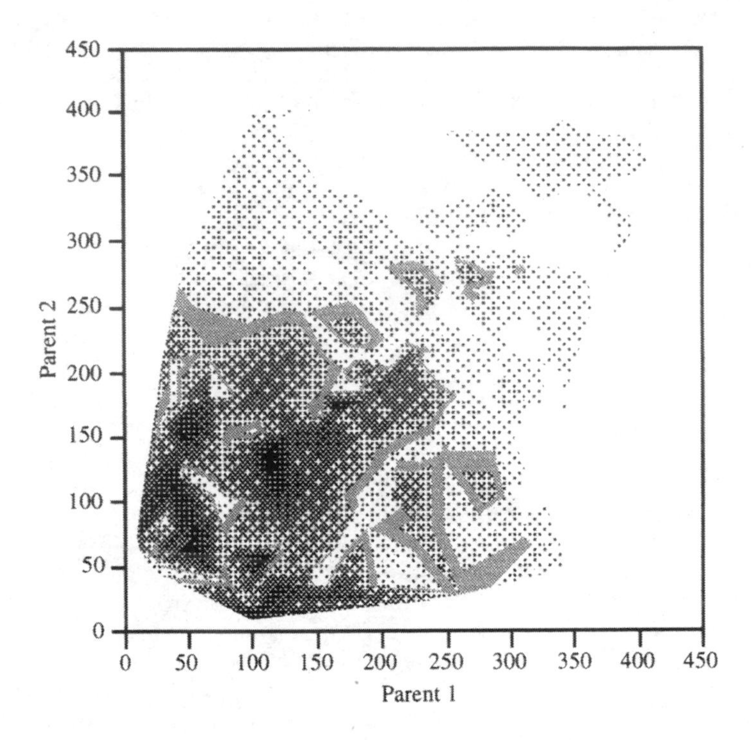

Figure 7.8 Plot of Gene Fitness vs Parents

The ability of GAs to 'sacrifice' good strings is exampled in figure 7.8. This plot comes about as a result of plotting parents against the fitness of their offspring, in a pure GA environment. The number of the parent is simply a number that represents a sequence of strings (chromosomes) as they are bred, therefore parent 100 was the one hundredth gene to be bred in this run of the model. (Note: dark areas indicate areas of low fitness and light areas indicate high fitness).

You will note that there are very specific bands of fitness and unfitness running through the generations, this is epistasis at work, and it is possible to trace good 'bloodlines' also you will note that as the generations are bred the general fitness of the new genes improve (become fitter). Occasionally an island of unfit or fit genes appear to be generated for no particular reason, this can be explained by the mutation of genes. These mutations can dramatically alter the gene structure and hence the fitness value of any gene that is mutated. Such an island of high unfitness seems to have been started by the breeding in the area of gene 125 parent1 and 125 parent2 and is slowly bred out also the breeding near 175 and 175. However; these unfit strains are very rapidly bred out.

You will also note that 'good' families disappear and then reappear. This can partially be explained by the survival of partial strings. These partial strings remain 'hidden' in a string and later combine with some other set of partial strings to create very fit offspring from a breeding cycle.

A Brief History of GAs

The breadth of applications that incorporate GAs is one of the more impressive aspects of this very young field of endeavor. Because of this wealth of material we shall, therefore, limit the review to important milestones in the short history of GAs.

Back in 1975 Professor John H. Holland, in his book "Adaptation in Natural and Artificial Systems", set forward the foundations for the genetic approach to search and optimization. The publication of his work marked the public birth of GAs. Professor Holland's book is a concise and coherent collection of his ideas, developed over a significant period of time before the book's final compilation and publication.

The professor started with a general study of adaptive systems during the early 1960s, he then proceeded to work on the role of recombination in adaptation (Holland, 1971), and the role of schema. Then, in 1975, Professor Holland was ready to join it all together and the book published in 1975 does that - joins it all together. Even thought there were significant contributions made by a number of other researchers (mainly students at the University of Michigan, the 'Mecca' of GAs for a long time) which bear on the theme of GAs, GAs are even now, justifiably, associated with Professor Holland and his pioneering work in the field.

Until the early 1980s, the research in GAs was mainly theoretical, as is required in the early development and construction of a new discipline, with very few real applications being considered let alone attempted. This period of time is partially dominated by work with fixed length binary representation in the domain of function optimization (De Jong, 1975). De Jong's work formalized the framework for simple GAs and much of the subsequent work in GAs followed in his footsteps.

The research into the workings of GAs and the development of theories of their operating mechanisms continued. After the slow appearance of real-world applications in the early 1980s GAs went on to experience a wide array of applications which spanned a number of disciplines. Every new application gave new perspectives and dimensions to the diversity and applicability of the discipline. Furthermore, in the process of improving performance as much as possible via tuning and specializing the GA operators, new and important findings regarding the generality, and robustness of GAs became available.

In engineering, Goldberg's work on steady-state and transient optimization of a gas pipeline using GAs is a classic in the field (Goldberg, 1983). There are other notable works in engineering (Davis and Coombs, 1987; Fourman, 1985; Glover, 1987; Goldberg and Samtani, 1986), pattern recognition (Englander, 1985; Grefenstette and Fitzpatrick, 1985; Stadnyk, 1987; Wilson, 1985), neural networks implementations (Ackley, 1985; Cohoon, et al., 1987; Dolan and Dyer, 1987; Jog and Van Gucht, 1987; Suh and Gucht, 1987; Tanese, 1987). Other types of applications have covered a wide area of endeavor, from the configuration of stack filters (Chu 1989), the design of aircraft (Bramlette & Cusic 1989), picking winners at the race track (Maza 1989), determination of optimal behaviours for Oligopolists (Marks 1989), chemometrics (Lucasius & Kateman 1989), scheduling the release of work into a flow shop (Cleveland & Smith 1989) and solving the traveling salesperson problem (Jog, Suh & Gucht 1989).

The field has continued to grow and still offers a wealth of unexplored ground for researchers to follow. It has been one of the more useful and applicable 'discoveries' of the last few decades and continues to prove to be of value in real-world applications. As a field of endeavor it continues to help in solving problems that until a few years ago would have been considered intractable or not worth the resources that would have been required to solve then.

8 Where we Need to Go from Here

Carver Mead, visionary physicist, Professor of Computer Science (California), on Robert Noyce's invention, in 1959, of the integrated circuit:

It came from a very, very stupid question about something we were doing that was even more stupid'.

Business Review Weekly, April 8, 1988.

Conceptual Modelling

We believe the future of modelling is in this area, the area of Conceptual Modelling (system dynamic modelling). Models to date have generally been of the data in - results out type. However; there is another type of model ideas/concepts/beliefs in - foresight out.

We, in the transport arena, use only the first type of model; the conceptual models being left to those that deal with high degrees of uncertainty (usually in the social science areas). Detailed raw data is the input to our models and concrete, reproducible and quantitative results the output. If we, as a profession, are concerned more with strategic/policy questions then we have a problem since it is not possible to obtain hard and fast data on these types of questions.

Strategic policy formulation is an amorphous undertaking. We do not have clear cut answers to clear cut problems in many situations. We therefore need to develop systems that allow for the input of broad concepts, ideas and beliefs. Output from such a model defines 'pictures' of the future environment, given these ideas and beliefs. From these pictures we can determine remedial actions that need to be undertaken to either avoid the fulfillment of an inimical predicted future or to cause a preferred future to come about; whichever is required.

These models do exist (the 'World Model' [Club of Rome] was one such, Genie is another). Models that require cross-price elasticities or accurately formulated equations as inputs and computational mechanisms need extensive amounts of up-front detailed work and research to obtain the inputs and equations and even then the accuracy of such data/equations is questionable. However; a statement such as; 'Vehicle usage will decline if the price of fuel increases' is a universally accepted truism and is not inaccurate nor will it change over time. Therefore we can define a model that will accept the following type of operation: If the price of fuel increases the amount of vehicle usage will decrease. The magnitude of the effect can be one of a number of simple choices; high, medium or low (ie the change is either greater than, equal to or less, but still in the same direction, than the movement in the affecting factor).

If we can develop systems that require inputs of the form described above then we can build robust, useful and easy to use models. There needs to be a change in the way we attempt to define and react to proposed futures and the tools to assist in that definition are some of the first things that need to change.

We loose nothing by trying and potentially have much to gain. I believe that these types of models can answer a number of questions that we cannot answer in any other manner and I also believe that these answers are a necessary prerequisite to the solution of many transport problems.

We need to define controls (mechanisms for the inputting of our concepts of effect between factors). Below (figures 8.1 and 8.2) are illustrated three possible representations of input controls.

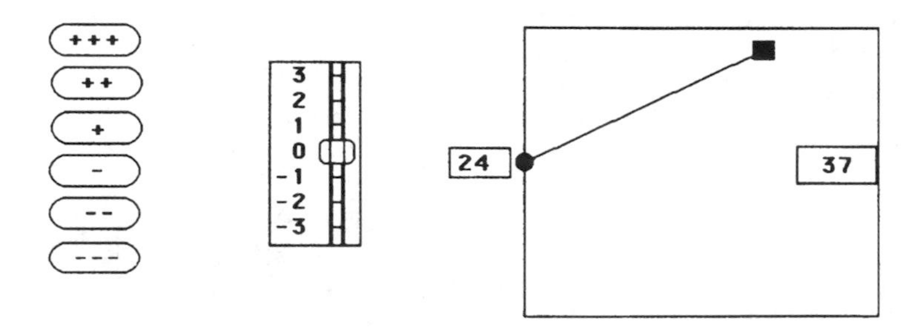

Figure 8.1 Possible One and Two Dimensional Control Types

In the first we click on a button to define whether the effect is highly positive (three plus signs) or slightly positive (one plus sign). We can also define negative effects in the same manner. In the second we drag the button up or down the scale to input the same types of effects. In the third we drag the black box on the end of the line. The angle, positive or negative can define the degree of positive or negative effect and the length of the line can be employed to input the magnitude of some other effect.

It would be just as simple to define three factor effects using a three dimensional analogy.

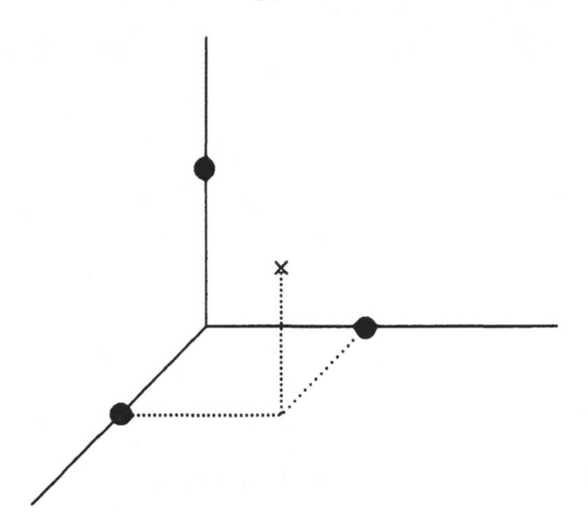

Figure 8.2 Possible Three Dimensional Control

The diagram above is a three effect control. The input is developed by dragging the black dots along the relevant axis. If more than three effect magnitudes were required then a new paradigm would need to be developed.

The strength of these types of input mechanisms is that no direct correlation between the input levels and hard data is assumed. Effect inputs can be relative as opposed to absolute (eg I believe that the magnitude of factor 1's effect is greater than factor 2's effect on factor 3 and can input these relative effects without recourse to extensive data collection or research). One can also make decisions on effects without relative evaluations (ie I believe that factor 4 has a very high negative effect on factor 5).

Genetic Algorithms

Genetic Algorithms are, in and of themselves, a fruitful area of research. The types of search problems they are best able to handle has not yet been quantified to any degree nor have the various reproductive mechanisms that can be employed on any particular type of problem been searched exhaustively.

The type of population characteristics employed in the genetic algorithm used in this book is what is termed a Panmictic Population.[50] There is however one slight difference from a pure Panmictic Population and this is the fact that there are no sexes described in the algorithm employed.

Following is a list of possible mechanisms that could be researched:

Dominance

Sexual Reproduction

Migration (immigration and emigration)

Selection of Reproducing Pair

Families (Within family only [form of inbreeding], outside family only [cross breeding] or either [covered by the general case])

Reproduction within classes only (Inbreeding) or outside classes only. Individual or family classing.

Scaling Mechanisms

Offspring replacement of parents or random replacement.

Steady population or increasing in size.

Limit the reproductive lifetime of a chromosome or unlimited reproductive life.

Within a family maintain parent reproduction or allow incest.

Choice of family for offspring. Based upon random selection independent of parental lineage, randomly from parent lineage or from a particular parents lineage. Offspring go to family with highest average fitness or lowest average fitness, based upon a patriarchal or matriarchal society or some other form of selection mechanism.

Do couples that 'marry' start new families or join a particular family?

Allow inbreeding only after there has been sufficient interbreeding, free inbreeding or no inbreeding at all, at any level.

Assortive Mating. Breeding between groups (negative assortive) or within groups (positive assortive).

---

[50]According to Elandt-Johnson (1971) pg. 60 a Panmictic Population has the following characteristics:

1) The fertilities and the survival abilities of each genotype and its gametes are the same for each individual in the population, and they remain the same through generations. In other words, the reproductive abilities (called fitness) of different genotypes and matings are the same and are not influenced by other forces such as selection, mutation, or migration. (This implies that, on the average, all mating types produce the same number of offspring.)
2) The ratio of females to maels [*sic*] is 1:1.
3) The generations are nonoverlaping.
4) The maternal and paternal contributions to the genotype (with respect to autosomal [sexless] loci) are the same.

Calculation of breeding values for all parents. Determined by the fitness of all that parents offspring.
Inbreeding ratio.
Determination of Genetic and Evolutionary loads (Ewens [1979]).
Determination of Optimal Starting Population Size (Goldberg [1989]).
Adaption of Operator Probabilities under Program Control (Davis [1989]).
Crossover Bias (Eshelman, Caruana & Schaffer [1989]).
Different Crossover Mechanisms (Syswerda [1989]).
Niche and Species Formation (see figure 8.3 and Deb & Goldberg [1989]).[51]

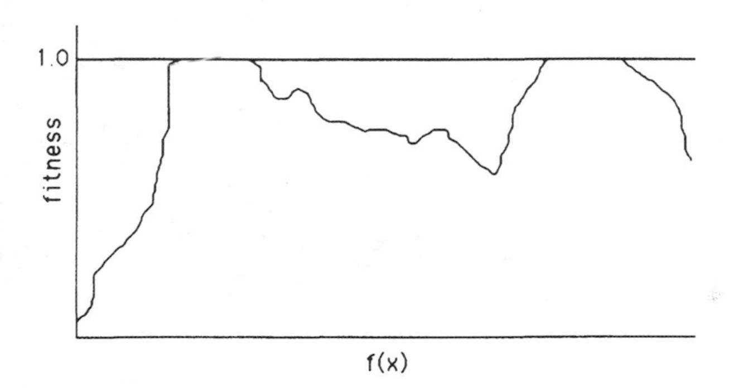

Figure 8.3 Example of Multi-optima Function

Currently the discussion on genetic algorithms has limited itself to 0-1 genes. However; there are situation in which we do not require the measurement of the magnitude of some value, by decoding a number of genes, but rather the presence or absence of some attribute (ie a pure 0-1 or on-off). Also there may be a need to handle a situation where we can make a choice between a number of states for a particular attribute (ie blue, green or red). In situations such as these we require a new mechanism for handling and manipulating a particular gene which may not be limited to a 0-1 state but rather an A, B, C or D type and further maybe states such as A, b, c or D where some form of dominance becomes apparent and can be manipulated and evaluated (decoded).

[51] In the application of the GA described in this book niche formation is very important. It is patently possible to have a number of potential solutions to the problem being analysed. In the simple GA implemented there is a tendency to converge to a single optima. There is a need to determine all potential solutions. The work by Deb and Goldberg (1989) describes methods of eliciting all possible optima and/or sub-optima in a single run.

Rules for the decoding and manipulation of multi-attribute genes are yet to be formalized and could prove to be a very fruitful area of research and further study. Also it must be remembered that there are possibly mechanisms of genetic manipulation available that do not reflect any operations available in nature. In the readings I have made there appears to be a tendency to dismiss any operations that do not mimic natural operations. I find this myopia rather disconcerting, the implication being that anything not found in nature has no value. I would contend that it is eminently possible to generate new mechanisms of manipulating the genetic code, to produce valid results, that have nothing in common with the manipulations performed in nature.

9 Paradigm Shifts - Trends and Coming Changes in Travel Patterns

> Vision is the art of seeing things invisible.
> - Jonathan Swift.

Current Trends and their Relationship to Current Travel Patterns

> There is no hope, but I may be wrong.
> - Bumper Sticker

Demographies of the Future

Smaller, more electronically aware, more scared of the external environment, more self-sufficient and more environmentally aware.

We are also seeing an increase in the proporation of aged in our societies - these people tend to take shorter trips and these trips are generally outside of peak periods. We can then expect a relative decline in peak and an increase in off-peak travel.

Hamish McRae (1994) when talking about changes in demographies had the following points to make that are relevant to this discussion:

- Retirement ages have to rise
- Female participation ratio in the workforce will climb
- Part-time working (including homeworking) will continue to rise
- Voluntary labour will be used to a greater extent and
- Greater efforts will be made to see the unemployed are in work.

Strategic Planning

Each of these factors will have a significant effect on travel patterns. Each, to some degree, will reduce the demand for peak period travel and will contribute to reductions in congestion. Each of these demographic groups will increase pressures for flexible hours and/or more home based work.

Cocooning, Cashing Out

The desire to avoid conflict and danger is causing more and more families to avoid the spaces outside their own homes and causes them to attempt, as much as is practical, to cocoon in their homes. Also the increasing tendency for people to leave the traditional work-place at an earlier and earlier age is also fueling dramatic changes to the travel patterns we are seeing evolve in some developed countries already.

The Electric Cottage

The introduction of electronic systems are allowing homes to become completely self-sufficient offices, shops (for ordering), schools and entertainment centres and are also improving the levels of service the home can offer in the more traditional areas such as health, and safety.

There has been much discussion over the last twenty years as to the changes that these systems will bring to demand for travel. There is a feeling that telecommunications will not create the changes first attributed to them. To date the figures banded about have not come to pass. In part this can be explained by a number of facts:

- The technology was expensive,
- The technology could not satisfy all needs of an office environment,
- The social pressures to avoid travel didn't exist,
- A critcal mass of applications and organisations available elctronically didn't exist,
- Managerial systems were not prepared to accept remote work,
- There were fewer jobs that could benefit from teleaccess and
- There were fewer organisations and cultures that could benefit from teleaccess as well.

Many of these impediments have disappeared and more are disappearing or reducing in relevance every day.

The Telecommunications Revolution

The sophisticated equipment that is required to maintain a dialogue with anyone, anywhere and anywhen is now so cheap and cost effective that it is more and more being found in the private home. Prohibitive costs ensured, until only recently that these levels of service delivery were only available to wealthy companies and were shared by a wide array of personnel. This cost ensured that such facilities were held centrally and if you required their use you had to be in "the office". Now you can be sunning yourself on Kuta beach and still be able to employ all these electronic services with ease. No longer do you need to be where the hardware is located - it moves with you.

Virtual Reality

"The real act of discovery consists not in finding new lands but in seeing with new eyes."
- Marcel Proust

More and more production systems are being managed, not by switches that need to be flicked manually, but rather by switches that are switched electronically. The question now arises - if the switch can be switched by an electrical signal then does the signal have to originate in a location close to the hardware? The answer is NO. It is possible for an operations 'room' to be located anywhere in the world. We can just as easily manage a nuclear power plant from Iceland as from the operations room of the plant in South Australia. This is true of all manufacturing, production and monitoring systems that are managed electronically.

There is an extention to VR that we think should be mentioned and that is stereolithography. This is a process in which data can be transformed into a physical object. Scans of a 3-D object are taken and stored on a computer. These data, that define the 3-D attributes of the object can then be transmitted to a manufacturing plant. At the plant, in a totally automatic process, can, from a bath of liquid plastic, create a duplicate of the 3-D artifact.

The obvious extension of this currently applied technology is that as the production systems become smaller and more compact, as the types of plastics that can be used become more varied that it would be possible for sites such as drilling rigs to start the manufacture of their own day-to-day drilling necessities, fabrication shops would manufacture their own car panels and bumper bars, retail outlets would produce furniture in their back rooms available for very rapid delivery to customers in the showroom.

The next step would be for the technology to end up in the home where you could create your own household non-consumables at the touch of a button. The only deliveries made to the home may then be consumables and a tonne of plastics for the households stereolithography production centre once a year.

Now tell us that travel patterns won't change!

And we still haven't discussed the possibilities and potentialities of nanotechnology and the changes it may bring about. Nanotechnology, if it fulfills only 10% of the rhetoric, could so dramatically alter the very fabric of our existance that the agricultural, industrial and information revolutions will appear trivial in comparison.

A Better Path

We are asleep with compasses in our hands.
- W.S. Merwin, American Poet.

Technology has and continues to define our urban environment. This can be proven by a simple review of the history of urban form. As transport became easier and easier throughout history so cities grew. Better transport systems made larger cities more 'profitable'.

The advent of the private motor vehicle made the corner store less 'profitable' than the supermarket. The corner stores mostly disappeared, supermarkets replaced them and we traveled more in our cars to get to those supermarkets.

The technology of the private car created freeways, parking bays, congested cities, health systems that struggle under the load created by motor vehicle accidents, contribute significantly to air and noise pollution and the significant geographic separations between home and work with all the attendant social and community dislocations associated with those seperations.

However there are new technologies that are becoming as ubiquitous as the private car and that offer even higher 'profits' for those that change to using these technologies. The nature of these systems to be more 'profitable' will ensure that they will, ultimately dominate the existing technologies - survival of the fittest works not just in biological systems but also in technological, economic, engineering and transport systems as well.

These technologies are electronic in nature - the personal computer, mobile faxes and phones, Internet, groupware, globalisation, intelligent agents, electronic entertainment, virtual travel, educational applications on screen, etc. and they are changing everything we do and how we do them. They are also creating new tasks and eliminating the need to undertake old ones.

Technologies are causing a fundamental paradigm shift in all the world cultures. The question for us is - how will these shifts affect transport?

The Energy Question

In 1987 proven oil reserves were 32.5 years' consumption. Were there to be no cutbacks in consumption and no new discoveries we would run out of oil in 2010/11 but this won't happen.

Long before the last drop of oil is pumped out of the ground the price will rise, conservation programs will be put in place, new discoveries will be made[52] or viable substitutes developed.

However; the use of free choice based transport that depends upon fossil fuels or its substitutes will be far more expensive than it is today and as a consequence choice based low valued travel will be reduced well before the year 2010/11 given all the available evidence.

Re-engineering

We believe that there needs to be a significant change in travel behaviours within our lifetimes. Urban planners, environmentalists, transport planners and the public at large seem again and again to say that this change needs to be fundamental in nature and that "fiddling about at the edges" won't fix the current or foreseeable problems.

What we are all saying, in so many words, is that the transport systems need to be re-engineered.

If we agree with this statement then we must also support the basic tenents of re-engineering which state, rather forcefully, that to achieve significant breakthroughs in levels of service and/or efficiency in the delivery of a product or service there needs to be "*fundamental rethinking* and *radical redesign* of ... processes to bring about *dramatic improvements* in performance."

Government Inability to Contribute

"No corporation gets hit by the future between
the eyes, they always get it in the temple."
- Dick Davis - Consultant.

[52] New discoveries will be relatively small. The world has been explored well enough for us to know that there exist no undiscovered fields like those in Alaska, the Middle East or Russia. New discoveries that are occuring today will only extend the lifetime of oil as a viable mass consumption energy source for a few years.

"Except Government - they simply loose a bit more credibility."
- addenda by the authors

Governments and their associated bureaucratic systems have historically proved to be well off the pace when it comes to offering any type of leadership in the preparatory phases of significant change. Governments react to rather than drive change, we don't expect the systems developed by government agencies, governments or political parties to deliver useful or sensible policies or initiatives in periods of rapid change. In the past they have themselves remained so patently unchanged and unaffected by the changes going on around them that they offer no hope of leadership in the turbulent times that are coming. Governments, of all political persuasions, are structured to manage steady state systems - in their early manifestations they may forment revolution to gain power but they always seek steady stable systems to govern once they have power. Governments cannot manage chaos and they are becoming less and less powerful and able to control complex systems in any case.

Experts?

"If God had intended that man should fly, He would have given him wings. The airship business is a 'fake', and will always remain so."
Rear Admiral George W Melville, US Navy, 1901.

The understandings and points of view of the experts in most fields are so entrenched that they cannot possibly drive a future they cannot see. As in nearly all disciplines we will probably find that it will be a non-transport person who forces us to change and to see the way things are going. Many of our ilk will be dragged screaming and kicking into the new world but it will happen - it is inescapable that the new technologies, the new social value sets, the new demands of the environment and of the next generation will alter the way we live, work, entertain and educate ourselves. This is inescapable. The only question is the form and function that this continually evolving system will take in the near future. We profer one possible future there are many others but each will have elements in common.

These common elements are those listed above, technology, social values, environment demands and the growing up of a new generation. We are able to forecast significant parts of these futures. Some of the trends are there and they are inescapable, some are not yet observable in any meaningful manner. However; all the systems we can expect to see emerging in the next twenty to thirty years will be developed from existing technologies. These we know and understand and we can describe how they are evolving and also some of their possible consequences.

Some of the Changes the Paradigm Shift will Bring About

Changes in Job Types - Infrastructure to Information

The trickle-down flow of Technology
First it goes to the military
Then to medical facilities
Then it's made into toys
Then to business
Then to education and
Finally to the general population (but with no instructions)
with apologies to Daniel Burrus

When our economies were dominated by production of physical goods there was a need to travel to observe, manipulate, purchase, deliver or alter these goods. There was no escaping the significant levels of demand for travel. This demand for travel continued to grow as our populations grew and as our demands for new and more physical products grew. Today the environment that determines the products and services we create has significantly changed that travel demand paradigm.

More and more our economy is dependent not upon physical goods but more and more on information and services. Goods that require to be physically transported are becoming less and less of the total transport task. Information can flow over wires and hence our concept of movement needs to move away from mobility in a physical sense and embrace movement in an electronic one as well. The electronic movement of the products of human labour will continue to grow and take a larger share of the total delivery task.

We generally believe there will be an ongoing increase in the levels of travel because we are not aware of any significant paradigm shift in the very near future. We believe there will be such a shift within the next two to three decades. This shift will see a dramatic alteration in the methods employed to observe, purchase and deliver goods and also their destinations.

Strategic Planning

The first three functions will be undertaken, for most goods, electronically and we also believe that mobility based home deliveries for physical goods will skyrocket.[53]

Overcoming of False Beliefs

Drexler (1992) and his co-authors postulate a list of false beliefs that have constrained development in a number of areas. We believe that the transport debate has been as hampered by these issues as have many of the other areas of human endeavour. These beliefs are:

- Industrial development is the only alternative to poverty
- Many people must work in factories (offices? question by authors)
- Greater wealth means greater resource consumption
- Logging, mining and fossil fuel burning must continue
- Manufacturing means pollution
- Third World development would doom the environment

These all depend on a more basis assumption:

Industry as we know it cannot be replaced.

Some further common assumptions:

- The twenty-first century will basically bring more of the same
- Today's economic trends will define tomorrow's problems
- Spaceflight will never be affordable for most people
- Forests will never grow beyond Earth
- More advanced medicine won't be able to keep people healthy
- Solar energy will never become very inexpensive
- Toxic wastes will never be gathered and eliminated
- Developed land will never be returned to wilderness
- There will never be weapons worse than nuclear weapons
- Pollution and resource depletion will eventually bring war or collapse

---

[53] Many electronic (eg computer software) or service based products (eg booking of airline flights) are already delivered directly to the home.

There is already evidence of a change in delivery practises for physical goods. Door-to-door parcel delivery services are one of the fastest growing industries in the world.

Vertical integration opportunites between retailers on the Internet and door-to-door parcel delivery are too obvious to pass up as an investment. In fact there is already evidence of these types of moves as well.

These too, depend on a more basic assumption:

Technology as we know it will never be replaced.

However; most of the real world indicators of industrial and corporate change would indicate that these stated basic assumptions no longer hold true. The shifts in power in our communities and public and private organisations being brought about by the rapid advances in technology show that industry as we know it is being replaced. The jury is still out on the question of a totally new form of technology. But at least the jury is deliberating the question.

When a technology/behaviour/system starts dominating a market it causes other alternatives to disappear even if the alternative selected is inferior.[54] The advent of the petrol car caused technological lock-in. We stopped looking for alternatives once we had the petrol/diesal driven car. The steam car may have been superior to the petrol driven vehicle but we got locked into petrol driven. The car brought about a paradigm shift that locked all else out. A paradigm shift not only displaces existing systems but also locks us into the new paradigm.

We believe that a set of existing beliefs on how the world works are ready to be displaced by a new set of beliefs. These new beliefs will be a new way of doing things - a new paradigm. And we'll get locked into this new paradigm just as we've been locked into all the others. The problem we confront is not will it happen but rather when and how can we ensure that the new paradigm is a good as it can be? Because once it catches hold we're stuck with it possibly for a long time along with all its good and bad points.

Some Social Commentary

Norman Myers (1990) paints a rather alarming picture of a possible future that we should not dismiss out of hand. As the global effects of the old-style destructive and selfish private car culture crashes headlong into the newer collective, caring and sustainable eco-culture that our children are growing up with we can expect significant clashes to occur. Can we expect

> outbursts of angry citizens against 'energy laggards'? For example, attacks on single-occupant cars and grosser gas guzzlers, on blantantly energy-inefficient buildings,

[54] VHS is an example of 'lock-in'. BetaMax is a superior system but since VHS got more customers in the first instance, it drew even more customers and became the 'standard' for video recorders. It wasn't the best, it just dominated the market. Another example is Macintosh vs. Wintel machines.

> or even fossil-fuel power plants. Will we see "energy wardens" authorised to issue on-the-spot fines to energy abusers? Will there be public pressure for governments to impose punitive carbon taxes on a host of energy-wasteful activities, both commercial and domestic?

We believe we can expect some of these behaviours to be manifest as time passes.[55]

Future Travel Patterns Arising from the Trends

If we do not change our direction, we are
likely to end up where we are going.
- Old Chinese Proverb.

All forms of teleaccess (telework, teleshopping, telelaw, telebanking, telecrime, telemedicine, teleinvesting, etc.) will bring about changes in travel patterns, collective delivery systems, the drive to cocoon, the capacity to be almost totally entertained and educated in your home, the social pressures not to drive a private vehicle that pollutes (air emissions, sound and safety)[56] and the need to deliver better, faster and lower priced products (goods and services) will see more of these alternatives being employed in the future.

With changes in social values and behaviours we can expect to see reductions in trips. These will be mostly for work, shopping, education and entertainment. These trip reductions will not be minor. We believe that within the next few decades trip reductions in the developed

[55] We have had significant confrontations in the last few decades because of cultural clashes and looking back on some of the reasons for those clashes we can laugh and now understand how silly they all really were.

Remember the 60's and 70's and long hair on men? Remember the fights between the fathers and sons? It was all because of a paradigm shift - a redefinition of what it meant to be masculine. Short back and sides simply couldn't handle the change in perceptions a new generation had. A relatively trivial reason for all the angst.

The paradigm shifts we are coming into are far more important than that - far more. We can expect a significant degree of social dislocation and friction. However; as has been the case through history the new paradigm, if viable and if it can answer some of the unanswered questions left behind by the old paradigm (which new accepted paradigms always can otherwise they never become mainstream), will win. The old paradigm will simply die away with its proponents as they age, lose power and authority and die off.

[56] We believe that safety, in a transport environment, is simply another form of pollution.

nations will be such that the air pollution and congestion seen in places like LA, Tokyo and London will be but memories of things past.

Some of the possibilities:

Work trips reduced by upwards of 60%: approx. 60% of workers are considered information workers all capable of telework. Further increases in the proportion of information workers could see this figure grow.

Education trip reductions of 75% - 90%: only hands-on laboratories will require physical attendance.[57]

Shopping trips reduced by 90% - 99%: only trips to purchase item such as fitted garments and others that absolutely require a physical presence will need to be undertaken.[58]

Entertainment trips will increase for destinations such as beach, picnics, etc. other types (movies, video store, games parlour, social clubs, etc.) will probably drop.

Aggregate Effects of Trends and Travel Pattern Changes

No matter how much you study the future, it will always surprise you; but you needn't be dumbfounded!
- Kenneth Boulding.

Within two to three decades we will wonder why we, in Perth, constructed a Northern City Bypass, extensive freeways and also why we bothered to extend the rail system; in other words why we spent hundreds of millions on infrastructure that could not be cashed in, sold off or used for alternative purposes when no longer needed to fulfil their primary function. The streets we already have will be more than sufficient to satisfy any potential population growth we could ever envision. New roads will be required but only in new developments but we need no new capacity to manage peak periods;

---

[57] We've had people say that this isn't possible since many families, because both parents work, require school as a child-care centre. However; if most people are working from home then the argument becomes mute. Some families will still not have parents at home and will require a form of child-care but the numbers will be nothing like they are now.

There is also the question of social interaction but this could be achieved by the development of localised play groups of kids living in the same street or close by.

[58] As technology improves the idea of what requires a physical presence will also change. The idea that it would be possible to see a set of plans without that set of plans being physically in the same room with you would have been laughable. Today with the communcations systems we have it is simple to see something, in detail, that resides physically a thousand kilometres away.

since traffic and travel patterns as we have today will simply not exist.[59]

Changing Work and Education Trip Patterns

The idea of a 9:00 - 5:00 workday will disappear since corporations will be required to work 24 hours a day 365 days a year to survive in a global economy. With the elimination of the 9:00 - 5:00 five day week the morning and afternoon peaks will disappear and with the disappearance of the peaks so disappears the need for broad, long and expensive freeways and much of the current infrastructure devoted to physical mobility.

The private car will NOT disappear it will simply be used in a very different manner. Travel to friends, parks, the beach, occasional shopping, face-to-face meetings, holiday travel, etc. will still occur. But we don't need massive city parking stations, large shopping centre carparks, freeways, bypasses, large and expensive public transport systems, schools and universities with significantly sized catchment areas, extensive policing of road users or hospitals larger than they need to be just to deal with road trauma.

It is possible today for any of us to, over the Internet, enrol in a degree at Harvard, Cambridge or Oxford without leaving home. We now seriously ask the question: 'If you can get a degree from any of these institutions why would you bother enrolling at your local university?'

One of the authors was told by a close friend that works with a WA university that there had been very high level discussions between a number of the universities in WA as to which campus would survive the coming changes since there was a belief that WA would possibly only require a single physical campus by the turn of the century.

It's not just universities that are vulnerable to these changes - schools (primary and secondary) are also vulnerable. When a single teacher can record a lesson that can be beamed to 50 million students as easily as 10 then there is the potential to decimate the worlds education industry. When 90% of students don't travel to school, along with all the other possible changes, we would contend that traffic patterns will be unrecognisable.

Economic and Workforce Changes

We will also see significant restructure in the workforce as these changes start to take effect. The motor industry is a significant part of

---

[59] Our traffic models of course do not show any of this since they are dependent on historic travel patterns - totally meaningless when the fundamentals change.

our economy. When our society consumes one tenth of the amount of oil we currently do, when we no longer need all the panel beaters, car yards, garages, RAC's and NRMA's, bus drivers, auto mechanics, transport planners, road builders and all the personnel that support these industries - what then? Also we need to realise that these changes will not just occur in WA but around the world. The total global effects are incalculable but will be significant, long-term and possibly socially devastating.

These same questions were last raised by our society just a decade ago with the advent of the desktop computer and the technologies that flowed from these (e.g. industrial robots, barcode readers).[60] We failed to plan for these changes in any meaningful and coherent manner. We believe the same will be true for any significant changes in travel patterns brought about by changes in behaviours and social values.

History has shown that the time it takes for a new social paradigm to be accepted and understood by a society is far greater that the time required for the paradigm to affect that society. By the time society and especially government recognises the change and prepares to act in a coherent manner it's too late - early adopters and change agents have created the atmosphere that will drive the shape, beliefs, understandings and values that the new shifts will bring.

A real horror is the fact that often the 'experts' in the field most affected by the paradigm shift are unaware, resistant, deprecating or outright saboteurs of these new systems even when the benefits of the system are obvious to all but the 'experts'.[61]

Future Trends

To develop trends of future private vehicle usage from current and past vehicle usage trends is farcical (see figure 9.1). Travel is a derived demand - find out what causes it, forecast changes in these behavioural drivers and then forecast travel demand. That's how it should be done but that's not how we do it.

That's why we believe many experts still believe that demand for physical travel, as opposed to virtual travel, will continue to increase.

---

[60] We may be very surprised about it all now but just 10 years ago unions attempted to prevent the introduction of barcode readers on electonic cash registers because of the forecasted job loses.

[61] A simple example was the introduction of Carbolic Acid as an antiseptic for use in surgery. The medical establishment fought against Lister for many years after he started to employ Carbolic Acid with tremendous success. Today our experts are not much different. The recent upheaval in the field of economics is a latterday example of the same mind sets dominating fields of endeavour as existed in Lister's time.

We forecast increases in population and without thought forecast increased physical travel as an almost linear function of population. This approach is trivial, morally corrupt, leads to expensive and incorrect decisions and is methodologically wrong.[62]

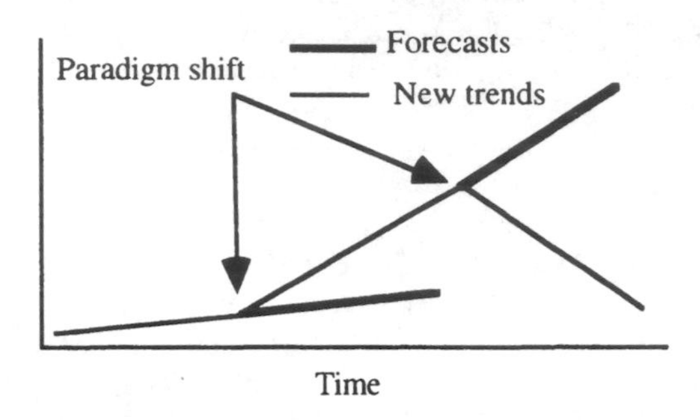

Figure 9.1 Paradigm Shifts

Some Simple Supporting Evidence

The National Travel Survey 1993/95 in the UK had the following statements to make regarding trips undertaken:

Pg 5; "fewer ... journeys than ... recorded in 1989/91"

Pg 21; "London residents travelled ... 22 per cent below the Great Britian average and 36 per cent less than other residents of the South East."

Pg 31; "the average number of commuting journeys ... fell by 14 per cent from 1975/76 to 1993/95"

"51 per cent of all commuting journeys began between 7 am and 9am or between 4 pm and 6 pm on a weekday. This is lower than earlier years and possibly reflects the spread of flexible working hour arrangements throughout Great Britian."

"fewer journeys are made by workers - down from a average for full-time workers of 442 to 389 in 1993/95."

---

62 It has taken a significant review of the systems employed to model forecasted travel demand in the UK to prove a number of methodological problems, the most important being that the transport models, generally employed around the world, incorporate a fudge factor that biased the results towards road construction. As was stated to me at a conference in Adelaide last year, "Don't tell me Traffic Models haven't been successful! They were constructed to ensure roads were built and by God they've succeeded."

In fact over the period 1975/76 to 1993/95 total worker trips (full and part-time) dropped from 776 per annum to 657 per annum a drop of 15 per cent.

Traffic and Transport as Complex Adaptive Systems

Many of us in transport would agree that traffic systems are Complex Adaptive Systems - it would be unlikely to encounter a question as to the truth of this statement - it is generally accepted as such.

Where we, as a profession, come unstuck, or so we believe, is in the execution of our models of traffic and transport systems. Our models are fundamentally linear in nature.[63] We run and continue to develop linear models because we believe we haven't the capacity to develop 'real' models. So we continue to make wrong decisions because many of us have been led to believe that this is the best we can do. We disagree!

We feel that the science of traffic engineering is much like the science of economics. Caught up in a form of self-consciousness and embarrassment for being somewhat less than 'real science' or 'real engineering'. As a profession therefore we try to cover-up our fear by being as rigorous as possible. We build mathematically correct models that describe systems in equilibrium; a place we know traffic never gets, and model little of the reality we can walk out of the door and see (much like the weather forecaster that tells you it's all clear when you are standing out in the rain). We know our forecasts are wrong, yet we continue to defend the indefensible. Just like neo-classical economists who when the real economy doesn't match the forecasts blame the economy for not 'doing the right thing' rather than blaming their unrepresentative models.

We leave many variables out of our models because we cannot obtain meaningful data, because we cannot forecast their future values with sufficient accuracy (sometimes we can't forecast them at all and sometimes they do go into models even so). We justify these omissions/additions by attempting to convince ourselves that these factors are not important, won't affect the end results significantly, or that their effects are swamped or dampened by others.

If we cannot model effectively then why do we not take another route? Attempt to understand traffic as a complex self-adaptive

[63] Complex systems are non-linear in nature and cannot be adequaely modelled by linear approximations. Adaptive complex systems, of which traffic is an example, are even more difficult to model adequately and yet we all agree that this is the reality we are confronting.

system. The advantage of this possibility is that it may be possible to 'understand' traffic by the development of a few simple rules.

Here we would like to move the discussion to cellular-automata. In 1984 Stephen Wolfram (Waldrop [1992], pg 225) determined that cellular-automata had a rich mathematical structure but have significant similarities to non-linear dynamic systems (like traffic systems). Wolfram also contended that there were four universality classes that all cellular-automata rules fell into. These classes were:

> Class I: contains what could be called doomsday rules. No matter what pattern of living or dead cells you started out with, everything would just die within one or two generations. In the language of dynamic systems the rules seemed to have a single 'point attractor'. That is, the rules seemed to mimic a mathematical surface that was shaped like a gravity well, the system would roll down the sides of the well and sit at the bottom, in a dead state.
>
> Class II: these rules were a little more lively, but not by much. With these rules the initial set up of living and dead cells would quickly coalesce into a set of static blobs, perhaps with a few other blobs that would periodically oscillate. These automata give the impression of stagnation and death. Here there existed a set of periodic attractors. A set of bumps in the mathematical surface within which the system could move indefinitely.
>
> Class III: these went to the opposite extreme, they were too lively. They produced so much action that the cells boiled with activity. Nothing was stable or predictable: structures would break up and reform almost immediately. Here were the 'strange attractors' so familiar in chaos. Here was a system that had been shot onto the mathematically surface with so much energy that they would never settle down.
>
> Class IV: here were the rare systems, the impossible to pigeon-hole rules that didn't produce frozen blobs, but that also failed to generate total chaos. What they did produce were coherent structures that grew, split, and recombined into new complex and 'living' organisms. They never settled down. They were not predictable, but not chaotic either, they were complex self-adaptive systems just like the Game-of-Life; the most famous computer based example of the implementation of a set of class IV rules.

Given the evidence; when flowing well, traffic fits perfectly into the class IV rule set because of the evidence that the patterns of traffic flows, movements, etc., are not 'dead' or on the edge of 'death' and are certainly not chaotic. Traffic is rather orderly, directed and meaningful and becomes chaotic or dead only at specific times. There is certainly no 'death', no chaos, but rather complex behaviours that continuously undergo change and adaptation. The movement is directed towards a goal (much like a glider in the Game of Life). There exists a set of rules, that we may be able to discover, that will allow us to 'understand' the behaviour more clearly.

The change from class IV to classes I, II or III, which traffic goes through at times, can be likened to phase changes. Water changes from ice (class I or II) to water (class IV) and then to steam (class III). In the study of water we understand what brings about the phase transition - we don't in traffic but we should.

Studies in phase transition in chaos have shown that as we approach the onset of chaos there is an increase in computation. This is understandable if we realise that one route on the road to chaos is via an increase in complexity. The opposite is also true. If we reduce computation we reduce complexity and we approach class I and II behaviours.

Is the secret of remaining within a class IV rule set the maintainance of the correct level of computation? Could we employ the idea of computation to measure the effectiveness, efficiency and usefulness of our road networks?

At first this question may appear trivial in the extreme. However; a closer look could prove to be of value. What, in a traffic sense, could be represented by computation? We believe we can talk about the need for the controller of a traffic entity (a vehicle, a pedestrian, etc.) to concentrate on the job could be considered as a possible measure of computation. In a case of grid-lock there is no need for a driver to think about the job of driving the car – minimal movement: minimal computation (class I and II), as we increase the efficient use of the network we increase congestion, speed, decrease inter-vehicular distances, accelerate faster and stop faster, each of these acts requires increased concentration and effort (increased computation). Chaos in traffic would be a state where the ability of drivers to maintain the levels of concentration required failed (as exampled by an accident).[64]

---

[64] We would conjecture that an accident event (class III) in traffic immediately leads to localised death of the system (immediate phase change from class III to class I or II). Complexity increases until we make the phase transition from complex self-adaptive to chaos. However; it is impossible for a traffic system to remain in a chaotic state. Traffic, locally, comes to a relative or complete halt (class I and II rule sets dominate) immediately after the class III phase transition occurs and the accident occurs.

What needs to happen is to start to seek out the class IV rules that govern traffic systems and we haven't undertaken that search in any meaningful or directed manner as yet. What we, in the profession, have done is to dissect traffic into roundabouts, traffic lights, platooning, acceleration curves, etc., etc., ad-nauseam.

We don't get an understanding of LIFE by understanding, in microscopic or atomic detail, a living cell in all its seperable parts. LIFE involves synergy that is far more than the simple addition of all knowledge of biological systems that we currently possess. So to with traffic. An understanding of traffic is far more than the addition of our knowledge of roundabouts, traffic lights, platooning, acceleration curves, etc., etc., ad-nauseam.

Maybe one of the major stumbling blocks is that many, or maybe most, of those in the area do believe that understanding can be gained by the simple addition of micro-based knowledge. That by learning more and more at the micro-level we will, in time, come to understand the whole. That is a belief system, and like religion cannot be logically argued - for or against. However; new branches of knowledge would suggest that we will never understand traffic, or any complex self-adaptive system, by a dissection of all the parts.

New work being carried out at the Santa Fe Institute (Waldrop, 1992) has tightened up a theory that the 'optimal' position for growth and adaptation, for all systems, lies at the edge of chaos. This central idea has been shown to be true in a number of computer simulations of complex systems. The reason we must simulate these systems is that there exists no central quantitatively proven theory of complexity.

We ask you to develop a hypothesis in your mind. Posit a simple single link road network, with one origin and one destination, with one thousand vehicles traveling from the origin to the destination, and all required to arrive before some stated point in time (morning rush hour), all travelling at the same speed in a simplified model. Now lets start to increase the complexity of the system. The numbers of links between the origin and destination increases, more nodes, the capacities of the links differ, vehicles travel at different speeds and so forth. The complexity of the system has increased, but so too has the efficiency of the system (from a real world perspective). Complexity has increased and so has efficiency and effectiveness. The necessary levels of computation have also increased. Decisions need to be made that were not needed in the simpler scenario. Now we need to determine the route as well as the time to leave the origin to arrive at

---

We have, in our policing efforts historically, attempted to reduce the occurence of these phase transition events from class IV (complex self-adaptive) to class III (chaos). We should, maybe, concentrate just as much effort in defining ways of creating the phase changes required to move us from class I and II (dead or slow systms) into class IV (complex self adaptive).

the destination at the required time. In the simple scenario we only needed to work out the destination leave time.

If you look at many other systems, both natural and man-made, you will discover that these 'rules' generally hold true - increased complexity implies increased efficiency and effectiveness. The manufacture of consumer goods, delivery of products to market, political systems (our two-party system has become bogged down in a two way fight with little imagination or creativity on either side. Japan and Italy have been able to effect dramatic change, maybe partially because of the levels of complexity inherent in their political systems), educational (better directed educational opportunites, more subjects, greater demands, etc., all act to improve the quality of the system) and legal (the legal system is NOT in any real terms complex - enough said) systems; they improve their ability to satisfy their primary function as the systems become more complex and computationally intense.

However; there is a level of complexity that traffic systems cannot go beyond to remain efficient. Attempt to move beyond this point and there is a phase change to chaos, immediately followed by death. Systems, to maximise performance, move towards *c* (where complexity is maximised) and also attempt to change the shape of the surface near *d* to make it difficult to slip from the edge of chaos[65] into death *e* (see figure 9.2 from Woodcock and Davis [1980], p. 86).

---

[65] We are willing to posit that robust complex systems have a different shaped surface at the edge of the catastrophic cusp. The surface, would probably, curve sharply upwards at the edge of the cusp (point *d*), making it far more difficult to fall over the edge into death. This would allow those systems to be more effective than most by making it possible for them to increase levels of computation well above others before going through the phase transition from class IV to class III and then immediately to class I or II.

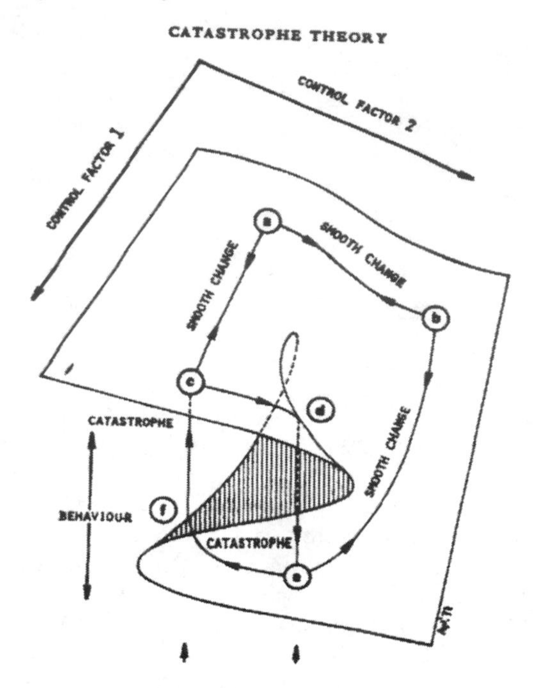

Figure 9.2 A Catastrophic Surface

One of the reasons, we believe that catastrophy theory has fallen into disfavour with the 'quantitative'[66] crowd is that it is not possible to use the theory to forecast exactly when an event will occur. It is however; very useful in helping us to understand the process and to maybe working out how to get out of the chaotic or dead zone once we get there. However; here again those of us caught up in the 'quantitative' approach are not particularly enamored by cerebral understanding without any quantitative support; we feel that that is unscientific and therefore unsupportive and therefore of no value. This view is, we believe, utter BS. Astronomy is undeniably science and yet an astronomer cannot forecast a Super Nova, they seek to understand the processes and not at a pure quantitative level. There are many sciences of like type. In transport we may benefit from the same perspective.

[66] Here we could just as easily have inserted 'Newtonian' or 'Reductionist' along with all the negative attributes associated with those terms today. As has been said 'we are today only just starting to recover from Newton.'

Let's try and develop a few class IV rules and understandings for traffic:

1) It is impossible to satisfy demand for road space in a sustainable fashion.
2) People will seek out the most flexible transport systems available and demand will cause that system to become cost-effective.
3) Value of time savings is a myth.
4) We will never be able to accurately forecast traffic behaviour or demand.
5) It is impossible to flatten out peaks sufficiently to eliminate the need for oversupply of infrastructure.
6) Public transport won't solve all the problems.
7) Getting the "Price Right" won't solve all the problems.
8) In fact, nothing will solve 'All the Problems'.
9) Building roads is simply like feeding a drug habit.
10) The private car is inevitable, in the short term, so we must learn to live with it rather than try to fight it head-on — we'll lose if we try.
11) Those with access to a private vehicle will use it.
12) It is very difficult to change driver behaviours.
13) Our existing modelling tools are a joke.[67]
etc.)

Support of New Knowledge and our Singular Lack of Success

As has been said "To climb the mountain one must first traverse the valley". Is it any wonder that established experts in traffic don't embrace a new approach; any new approach! If they are to reach the peak of a new mountain they first have to come down from the mountain top they currently dominate and traverse the valley with the rest of the beginners. No special position, respect or eminence in a new field of understanding. We all have to start again, but this time we risk much because just maybe we won't get to the top of the new mountain this time.

All experts, in any subject, sit atop a local optimum on the mathematical surface that describes their field. Knowledge continually grows and therefore the subject surface continually changes. If the new knowledge comes in at a controlled rate then it is possible for an existing expert to remain atop the subject. Learning causes the local

[67] Many of these are not rules but we thought they needed to be stated some place and this is as good as anywhere.

optimum upon which the expert sits to continue to grow and remain an optimum – of sorts. But what happens when new knowledge floods in, such as when there are totally new understandings and paradigms upon which we can operate? What then happens to the local optimum and the guru who sits atop it? It simply drowns and is 'metaphorically' washed away.

Think back through technological history. The names of certain persons still shine bright but their work is now considered technically trivial, uninspiring and everyday. Although their names float in the flood their work has sunk below the waves of change. The pity is when this happens to an individual that still lives or, worse yet, still seeks to steer or dominate a subject area.

The greatest tragedy is when those 'experts' are successful in their endeavours to maintain dominance by destroying the new seeds of knowledge because they are threatened in their position of eminence. It is for these that we reserve our greatest scorn and the greatest pity is that these people exist almost exclusively in our most respected institutions, places; supposedly; of intellectual search but, too often, of intellectual assasination.

10 Model Construction and Interpretation of Results

Global Warming?

The first model is one that was constructed to test the theory of a coming Ice Age as proposed by Prof Hamaker of Purdue University (note that any misinterpretation of his theory is purely the responsibility of the authors and in no way an be attributed to Professor Hamaker).

Table 10.1 is read as follows:

The factor descriptor is followed by a list of factors that are affected by the factor described. For example Vege Growth Rates will cause a reduction in Atmos CO2 levels and an increase in food supplies. The values given for each affected factor is a measure of the extent these factors are affected. A value of 0.5 implies that the affected factor moves at a rate of 50% of the movement rate of the affecting factor, therefore the model assumes that for a 100% increase in Vege Growth Rates there will be a decrease of 2% in Atmos CO2 and there will be a 3% increase in food supplies.

Table 10.1 Factor Description of the Ice Age Model

Description of Model Ice Age
No. of Factors: 14

1 Atmos CO2 levels		2 Vege Growth Rates	
Atmos CO2 levels	0.0010	Atmos CO2 levels	-0.0200
Vege Growth Rates	0.0300	Food Supplies	0.0300
World Temperature	0.0200		
Extreme Weather	0.0200		
Acid Rain	0.0100		

Factor	Value
3 Food Supplies	
Atmos CO2 levels	-0.0700
Population	0.0030
Birth Rate	0.0030
Rainfall	0.0300
Surface Water	-0.0030
4 Population	
Atmos CO2 levels	0.0500
Food Supplies	-0.0500
Population	0.0300
Birth Rate	-0.0030
Degredation of Soil	0.0700
Forest Fires	0.0200
Acid Rain	0.1000
Surface Water	-0.0030
5 Birth Rate	
Population	0.0050
6 Degredation of Soil	
Vege Growth Rates	-0.0600
Degredation of Soil	0.0100
Forest Fires	0.0600
7 Size of Artic Icecap	
Degredation of Soil	-0.0010
Size of Artic Icecap	0.0200
World Temperature	-0.0200
Volcanic Activity	0.0700
Extreme Weather	0.0400
Surface Water	-0.0500
8 Rainfall	
Vege Growth Rates	0.0200
Degredation of Soil	0.0050
Size of Artic Icecap	0.0200
Surface Water	0.0300
9 World Temperature	
Vege Growth Rates	0.0020
Size of Artic Icecap	0.0020
Rainfall	0.0300
Extreme Weather	0.0200
Forest Fires	0.0300
10 Volcanic Activity	
Atmos CO2 levels	0.0200
World Temperature	0.0010
Volcanic Activity	0.0200
Extreme Weather	0.0300
Forest Fires	0.0010
Acid Rain	0.0300
11 Extreme Weather	
Vege Growth Rates	-0.0030
Food Supplies	-0.0050
Size of Artic Icecap	0.0200
Rainfall	0.0020
World Temperature	-0.0020
Extreme Weather	0.0150
12 Forest Fires	
Atmos CO2 levels	0.0200
Extreme Weather	0.0100
Forest Fires	0.0100
13 Acid Rain	
Vege Growth Rates	-0.0500
Degredation of Soil	0.0200
Surface Water	-0.0020
14 Surface Water	
Rainfall	0.0400

The following results (table 10.2) show the changes that occur in the model factors over the period of the model run.

Table 10.2 Forecasts from Ice Age Model Run

Period	Atmos CO2 levels	Vege Growth Rates	Food Supplies	Population	Birth Rate	Degredation of Soil	Size of Artic Icecap
1	1.000	1.000	1.000	1.030	1.000	1.012	1.020
5	1.017	0.997	0.993	1.159	1.000	1.063	1.106
10	1.035	0.993	0.984	1.344	0.999	1.130	1.223
20	1.078	0.983	0.960	1.806	0.998	1.283	1.494
30	1.132	0.971	0.929	2.427	0.996	1.464	1.826
40	1.200	0.957	0.886	3.261	0.993	1.679	2.232
50	1.287	0.939	0.830	4.383	0.990	1.939	2.728
60	1.397	0.918	0.754	5.890	0.985	2.255	3.333
70	1.541	0.892	0.653	7.915	0.979	2.643	4.072
80	1.727	0.860	0.517	10.637	0.971	3.123	4.975
90	1.972	0.820	0.336	14.294	0.959	3.724	6.078
100	2.292	0.769	0.092	19.209	0.944	4.482	7.425
Period	Rainfall	World Tempurature	Volcanic Activity	Extreme Weather	Forest Fires	Acid Rain	Surface Water
1	1.000	1.000	1.021	1.017	1.011	1.000	1.000
5	1.000	0.998	1.112	1.085	1.057	1.015	0.995
10	0.999	0.996	1.236	1.178	1.119	1.036	0.989
20	0.998	0.991	1.527	1.389	1.253	1.089	0.975
30	0.997	0.985	1.886	1.638	1.406	1.160	0.957
40	0.995	0.977	2.330	1.933	1.580	1.253	0.935
50	0.993	0.968	2.878	2.283	1.780	1.377	0.907
60	0.990	0.956	3.553	2.697	2.011	1.542	0.873
70	0.986	0.942	4.387	3.189	2.279	1.762	0.831
80	0.980	0.925	5.416	3.772	2.593	2.055	0.779
90	0.973	0.904	6.686	4.466	2.963	2.445	0.715
100	0.963	0.878	8.252	5.290	3.401	2.966	0.635

The following graphs (figure 10.1) show (not to scale) the general directions of movement for the various factors.

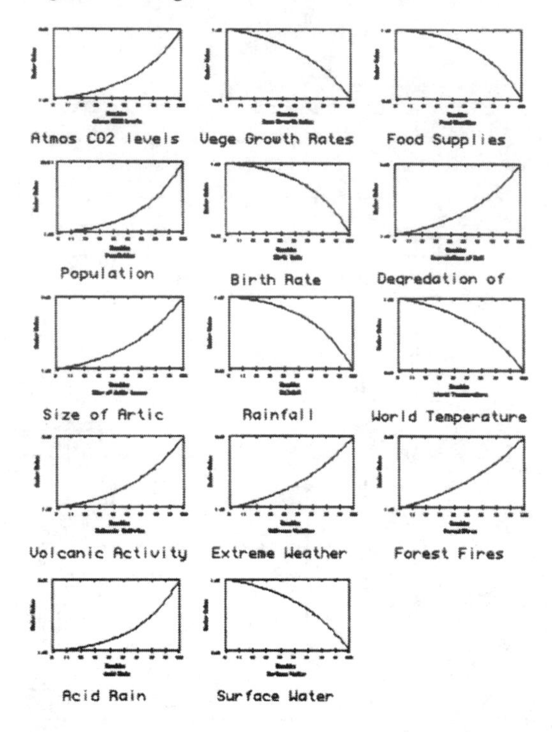

Figure 10.1 Factor Movements (not to scale) [68]

(Note: the x-axis indicates time and the y-axis the level of effect of the larger labelled factor. This is true for all graphs of this type in this appendix).

68 Figures 10.1, 10.2, 10.3 qnd 10.4 are output by the software package 'Genie' and cannot be alterered. The title for the sixth plot should be 'Degredation of Soil' in each of the three graphs.

After model GA run.
The task given to the GA was to maintain world tempurature, the "best" (closest result achievable) was selected for display (figure 10.2). As you will note, the temperature was almost maintained, but at what cost?

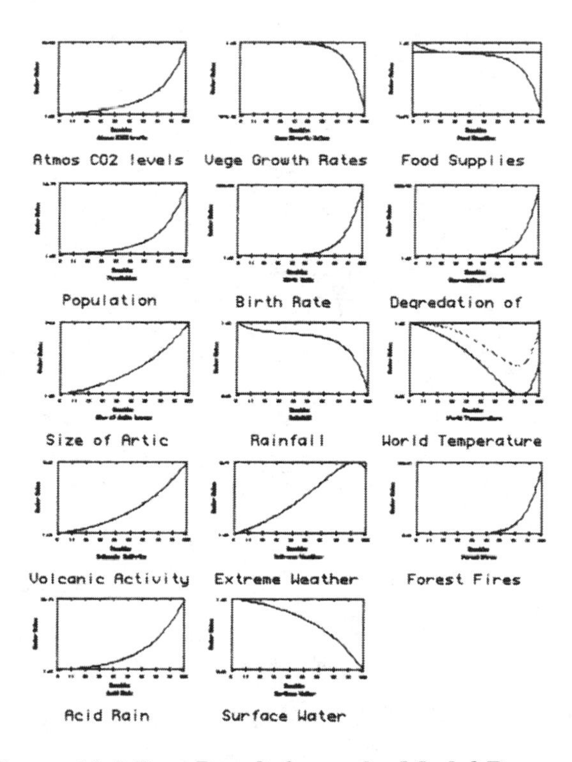

Figure 10.2 Best Result from the Model Runs

Table 10.2 supplies the actual raw data. As you will see the GA has selected a future that although possible is not preferrable.

Table 10.3 'Best' Model Results

Period	Atmos	Vege Growth	Food	Population	Birth	Degredation	Size of
	CO2 levels	Rates	Supplies		Rate	of Soil	Artic Icecap
1	1.023	1.000	0.905	1.030	1.083	1.096	1.020
5	1.118	0.968	0.602	1.161	1.487	1.577	1.105
10	1.237	0.911	0.358	1.348	2.211	2.481	1.222
20	1.498	0.690	0.112	1.826	4.892	6.101	1.494
30	1.810	0.174	0.010	2.488	10.829	14.929	1.826
40	2.197	-1.054	-0.050	3.417	23.977	36.436	2.231
50	2.693	-3.996	-0.117	4.755	53.090	88.792	2.727
60	3.360	-11.089	-0.239	6.753	117.557	216.196	3.332
70	4.322	-28.248	-0.506	9.877	260.312	526.154	4.070
80	5.850	-69.846	-1.123	15.049	576.426	1280.148	4.972
90	8.561	-170.821	-2.581	24.156	1276.427	3114.154	6.072
100	13.901	-416.124	-6.070	41.167	2826.509	7574.975	7.411
Period	Rainfall	World	Volcanic	Extreme	Forest	Acid	Surface
		Tempurature	Activity	Weather	Fires	Rain	Water
1	1.000	1.000	1.021	1.016	0.950	1.033	1.000
5	0.990	1.000	1.112	1.084	0.783	1.195	0.996
10	0.982	0.999	1.236	1.176	0.638	1.431	0.990
20	0.973	0.998	1.527	1.386	0.541	2.050	0.975
30	0.970	0.997	1.886	1.635	0.742	2.931	0.955
40	0.967	0.995	2.330	1.929	1.478	4.183	0.931
50	0.965	0.993	2.877	2.271	3.403	5.961	0.900
60	0.961	0.990	3.553	2.661	8.157	8.488	0.861
70	0.952	0.988	4.387	3.083	19.759	12.083	0.812
80	0.934	0.986	5.416	3.483	47.991	17.211	0.749
90	0.891	0.986	6.686	3.711	116.658	24.557	0.669
100	0.793	0.993	8.251	3.381	283.657	35.165	0.563

The next set of graphs are from another solution as recommended by the GA. Remember that the graphs are not to scale and that the world temperature remained within 20% of what was desired. The graphs are presented to illustrate the significant different types of recommended alternative that can be presented by the GA.

There has been a reduction in the recommended birth rate and in the setting of forest fires, these two recommendations have helped reduce the pressure on the weather systems and have helped maintain world temperatures.

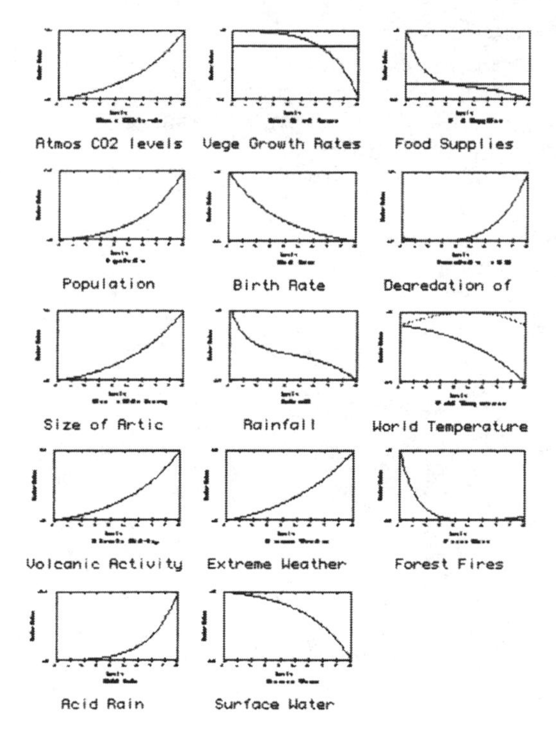

Figure 10.3 A Preferred Solution to the Ice Age Model

In the following examples of the test of the Hameker theory we would like to present the following.

In these tests we, as in the earlier examples, allowed the GA to modify the levels of Atmospheric CO2, Food Production, the Birth Rate, Degredation of Soil, Forest Fires, and Acid Rain. However, in the earlier examples we allowed the GA to determine the level of these factors to achieve its objective or maintaining world temperature in this example we set certain limits on the GA. These limits were that food production could not be reduced and that none of the other factors could be increased, in other words we are constraining the factors to moving in directions that we believe people would state were the preferred directions of movement.

We have selected the 'best' and the 'worst' that the GA had to offer after 20 generation runs.

The first gave a fitness of 0.057, maintaining world temperature rather well, the suggested changes to the controllable factors were:

CO_2	-0.057	reduce acid rain
Food	0.052	increase food production
Birth	0.000	maintain the existing birth rate (remember we constrained the birth rate so that it would not go positive. It looks like it went as high as the constraint would allow and may have preferred to go positive).
Soil	-0.087	maintain soil fertility. Expend significant effort to do so
Fires	-0.029	reduce burn-off, but not dramatically
Rain	-0.055	reduce acid rain

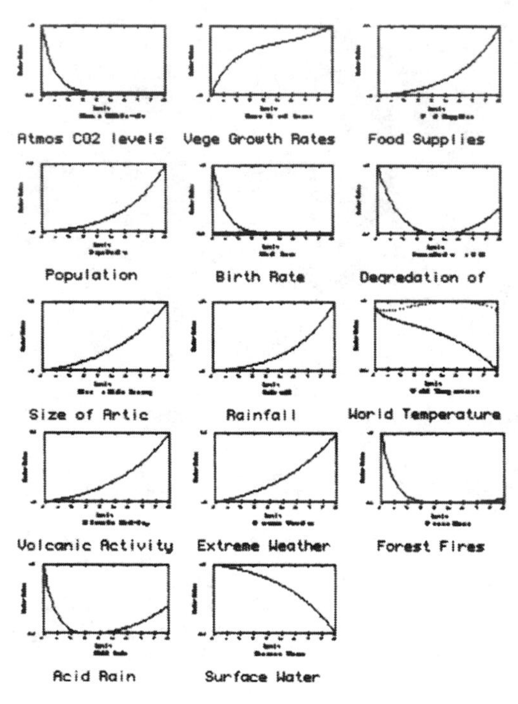

Figure 10.4 A Final Possible Solution

The second example achieved a fitness of 0.143, not nearly as good as the previous example, but still not an ice age. The GA suggested the following factor changes:

CO_2	-0.1	reduce CO2 generation dramatically
Food	0.031	increase food production

Birth	-0.1	reduce the birth rate dramatically
Soil	-0.049	reduce soil degradation
Fires	-0.094	significantly reduce burn-off
Rain	-0.098	reduce acid rain dramatically

The degree of factor changes suggested by the GA are those that we feel sure would be generally believed are required to 'save the world' however, the model suggests that if these extreme measures are taken then we end up in a worse state than if we take less extreme measures.

References

Ackley., D. H. (1985). A connectionist algorithm for genetic search. 1st International Conference on Genetic Algorithms, pp. 121-135.

Allison., G. (1971). *Essence of Decision: Explaining the Cuban Missile Crisis,* Little, Brown and Co., Boston.

Apostel., L. (1961). 'Towards the formal study of models in the non-formal sciences', in *The Concept and Role of the Model in Mathematics and Natural Science*, H. Freudenthal (Ed), Dortrecht, Holland.

Armstrong., J. S. (1978). *Long-Range Forecasting from Crystal Ball to Computer*. John-Wiley & Sons., New York.

Asada., H. and J. J. E. Slotine (1986). *Robot Analysis and Control.* John Wiley. N.Y.

Ashby., W. R. (1970). *An Introduction to Cybernetics*. Chapman & Hall Ltd. and University Paperbacks, London, reprint.

Atkins., S. T. (1987). "The Crisis for Transportation Planning Modelling." Transport Reviews. Vol 7 No. 4. Oct/Dec. pp. 307-325.

Barker., J. A. (1992). *Paradigms: Understanding the Future in Business and Life*. The Business Library, Melbourne.

Barker., S. F. and P. Achinstein (1971). "On the New Riddle of Induction", ch. X in P. H. Nidditch (Ed), *The Philosophy of Science*. London, Oxford Univ. Press.

Barnett., C. (1995). *Cyber Business: Mindsets for a Wired Age*. John Wiley & Sons, Chichester.

Batty., M. (1970). 'Recent developments in land use modelling: A review of British research'. Urban Systems Research Unit, Department of Geography, - University of Reading.

Batty., M. (1970). 'Models and projections of the space-economy'. Town Planning Review, vol. 41, 2.

Batty., M. (1970). 'Introductory model-building problems for urban and regional planning'. Urban Systems Research Unit, Department of Geography, University of Reading.

Bickel., A. S. and R. W. Bickel (1987). Tree structured rules in genetic algorithms. 2nd International Conference on Genetic Algorithms, pp. 77-81.

Belasco., J. A. and R. C. Stayer (1993). *Flight of the Buffalo.* Warner Books, N.Y.

Boulding., K. E. (1964). *The Meaning of the Twentieth Century: The Great Transition,* Harper and Row, N.Y. p. 80.

Boulding., E. (1973). 'Futurology and the Capacity of the West' in F. Tugwell. (Ed) *Search for Alternatives Public Policy and the Study of the Future,* Winthrop, Cambridge.

Bramlette., M. F. and R. Cusic (1989). 'A Comparative Evaluation of Search Methods Applied to Parametric Design of Aircraft', 3nd International Conference on Genetic Algorithms, pp. 213-218.

Brewer., G. D. (1975). 'An analysis of complex systems', in T. R. La Porte (Ed.), *Organized Social Complexity - Challenge to Polities and Policy,* Princeton Univ. Press, Princeton.

Broadbent., T. A. (1970). 'An urban planning model: what does it look like?' Architectural Design, August, pp . 408-410.

Brzezinski., Z. (1970). *Between Two Ages: America's Role in the Technetronic Era.* Viking, N.Y. pg. 76.

Burrus., D. and R. Gittines (1993). *Technotrends.* Bookman Press, Melbourne.

Byham., W. C. and J. Cox (1988), *Zapp! The Lightning of Empowerment.* Fawcett, N.Y.

CAMBRIDGE UNIVERSITY. (1970). 'Cambridge: the evaluation of urban structure plans'. Land Use and Built Form Studies, WP 14. Cambridge, July.

Chadwick., G. (1971). *A Systems View of Planning.* Pergamon Press, Oxford.

Chambers., L. D. and M. A. P. Taylor (1991). 'Where are we going?: the use of the trends integration procedure (TIP) for scenario generation', Traffic Engineering & Control. Forethcoming.

Chu., C-H. H. (1989). 'A Genetic Algorithm Approach to the Configuration of Stack Filters', 3nd International Conference on Genetic Algorithms, pp. 219-224.

Clark., J and S. Cole (1976). *Global Simulation Models.* Wiley - Interscience, London, (see pp. 110-112).

Cleveland., G. A. and S. F. Smith (1989). 'Using Genetic Algorithms to Schedule Flow Shop Releases', 3nd International Conference on Genetic Algorithms, pp. 160-169.

Coates., J. F. (1985). 'The Future is Our Business', Fifth Annual Report of the Commission of the Future.

Cohoon., J. P., et al. (1987). 'Punctuated equilibria: A parallel genetic algorithm', 2nd International Conference on Genetic Algorithms, pp. 148-154.

Cordey-Hayes., M. (1970). 'Structure plans and models', Architectural Design, July, pp. 362-363.

Cuypers., J. C. M. and O. Rademaker (1973). 'An Analysis of Forresters' 'World Dynamics' Model', Project Globale Dynamica, Report U16, (1973-05-01), Eindhoven.

Davidow., W. H. and M. S. Malone (1992). *The Virtual Corporation.* HarperBusiness, N.Y.

Davies., P. and J. Gribbin (1991). *The Matter Myth.* Penguin, London.

Davis., L. and S. Coombs (1987). 'Genetic algorithms and communication link speed design: Theoretical considerations', 2nd International Conference on Genetic Algorithms, pp. 252-256.

Davis., L. (1989). 'Adapting Operator Probabilities in Genetic Algorithms', 3rd International Conference on Genetic Algorithms, pp. 61-69.

Dawkins., R. (1986). *The Blind Watchmaker.* Longmans Scientific & Technical, Harlow.

De Jong., K. (1975). An analysis of the behavior of a class of genetic adaptive systems (Doctoral dissertation, University of Michigan). UMI Dissertation Information Service.

Deb., K and D. E. Goldberg (1989). 'An Investigation of Niche and Species Formation in Genetic Function Optimization', 3rd International Conference on Genetic Algorithms, pp. 42-50.

Dolan., C. P. and Dyer, M. G. (1987). 'Towards the evolution of symbols', 2nd International Conference on Genetic Algorithms, pp. 123-131.

Drexler., K. E., C. Peterson and G. Pergamit (1992). *Unbounding the Future.* Simon and Shuster, London.

Drexler., K. E. (1992). *Engines of Creation.* OUP, Oxford.

Drucker., P. (1969). *The Age of Discontinuity,* Harper and Row, N.Y. pg. 194.

Dyson., F. (1988). *Infinite in all Directions.* Penguin, London.

Elandt-Johnson., R. C. (1971). *Probability Models and Statistical Methods in Genetics.* John Wiley & Sons, N.Y.

Ellyard., P. (1991). Seminar of Apple Computers (Aust) 'OpinionMakers' series at the Hyatt Hotel, Perth, 20th June.

Englander., A. C. (1985). 'Machine learning of visual recognition using genetic algorithms', 1st International Conference on Genetic Algorithms, pp. 197-202.

Eshelman., L. J, R. A. Caruana and J. D. Schaffer (1989). 'Biases in the Crossover Landscape', 3rd International Conference on Genetic Algorithms, pp. 10-19.

Ewens., W. J. (1979). *Mathematical Population Genetics.* Springer-Verlag, Berlin.

Forrester., J. W. (1971). *World Dynamics,* MIT Press, Cambridge, Massachusetts.

Fourman., M. P. (1985). 'Compaction of symbolic layout using genetic algorithms', 1st International Conference on Genetic Algorithms, pp. 141-153.

Gleick., J. (1987). *Chaos,* Cardinal, London

Glover., D. E. (1987). ‚Solving a complex keyboard configuration problem through generalized adaptive search', In L. Davis (Ed), Genetic algorithms and simulated annealing (pp. 12-31). Pitman , London.

Godet., M. (1986). 'Introduction to La Prospective'. Futures, 18, 2, 1986, pp. 134-157.

Goldberg., D. E. (1983). Computer-aided gas pipeline operation using genetic algorithms and rule learning (Doctoral dissertation, University of Michigan). UMI Dissertation Information Service.

Goldberg., D. E. (1989). *Genetic Algorithms in Search, Optimization & Machine Learning*. Addison-Wesley, Reading.

Goldberg., D. E. (1989a). 'Zen and the art of genetic algorithms', 3rd International Conference on Genetic Algorithms, pp. 80-85.

Goldberg., D. E. (1989b). 'Sizing Populations for Serial and Parallel Genetic Algorithms', 3rd International Conference on Genetic Algorithms, pp. 70-79.

Goldberg., D. E. and M. P. Samtani (1986). 'Engineering optimization via genetic algorithm', 9th Conference on Electronic Computation, pp. 471-482.

Goldberg., D. E. and R. E. Smith (1987). 'Nonstationary function optimization using genetic algorithms with dominance and diploidy', 2nd International Conference on Genetic Algorithms, pp. 59-68.

Goldberg., M. H. (1990). *The Book of Lies*. Quill/William Morrow, N.Y.

Goldratt., E. M. and J. Cox (1986). *The Goal: A Process of Ongoing Improvement*. North River Press, N.Y.

Greene., D. P. and S. F. Smith (1987). 'A genetic system for learning models of consumer choice', 2nd International Conference on Genetic Algorithms, pp. 217-223.

Grefenstette., J. J. and J. M. Fitzpatrick (1985). 'Genetic search with approximate function evaluations', 1st International Conference on Genetic Algorithms, pp. 112-120.

Hagget., P. and R. J. Chorley (1967). 'Models, paradigms and the new geography', in *Models in Geography*. London, pp. 19-41.

Hammer., M. and S. A. Stanton (1995). *The Reengineering Revolution*. HarperBusiness, Sydney.

Hilmer., F. (1990). *New Games New Rules*. Angus & Robertson, North Ryde.

Holland., J. H. (1971). Processing and processors for schemata. In E. L. Jacks (Ed.), *Associative Information Processing* (pp. 127-146). American Elsevier, N.Y.

Holland., J. H. (1975). *Adaptation in Natural and Artificial Systems*. University of Michigan Press, Ann Arbor.

Holland., J. H. and J. S. Reitman (1978). Cognitive systems based on adaptive algorithms. In D. A. Waterman and F. Hayes-Roth (Ed.), *Pattern Directed Inference Systems* (313-329). Academic Press, N.Y.

Hutton., D. W. (1994). *The Change Agent's Handbook*. ASQC Quality Press, Milwaukee.

Ingwell., F. (Ed.) (1973). Search for Alternatives: Public Policy and the Study of the Future, Winthrop, Cambridge. pg. vi.

Janis., I. L. and L. Mann 'Coping with decisional stress', American Scientist, Vol. 64, pp. 657-67.

Jog., P. and D. Van Gucht (1987). 'Parallelisation of probabilistic sequential search algorithms', 2nd International Conference on Genetic Algorithms, pp. 170-176.

Jog., P., J. Y. Suh and D. Van Gucht (1989). 'The Effects of Population Size, Heuristic Crossover and Local Improvement on a Genetic Algorithm for the Traveling Salesman Problem', 3nd International Conference on Genetic Algorithms, pp. 110-115.

Johnson., J. H. (1983). 'The Logic of Speculative Discourse: Time, Prediction, and Strategic Planning', Environment and Planning - B, vol 9, pp. 269-294.

Levy., S. (1992) *Artificial Life*. Penguin, London.

Lippmann., W. (1966). 'Today and Tomorrow-Catching up with the times', Washington Post, 14 November.

Lucasius., C. B. and G. Kateman (1989). 'Application of Genetic Algorithms in Chemometrics', 3nd International Conference on Genetic Algorithms, pp. 170-176.

Marks., R. E.. (1989). 'Breeding Hybrid Strategies: Optimal Behavior for Oligopolists', 3nd International Conference on Genetic Algorithms, pp. 198-207.

Markeley., O. W. (1983). 'Preparing for the Professional Futures Field'. Futures, Feb, pg. 47.

Marris., P. (1974), *Loss and Change*, Pantheon, N.Y. p. 170.

Martino., J. P. (1972a, b, c, d). *Technological Forecasting for Decision-making*. American Elsevier, NY. (a: pg. 67)(c: pg. 130)(d: pg. 168).

Maza., M. de la. (1989). 'A SEAGUL Visits the Race Track', 3nd International Conference on Genetic Algorithms, pp. 208-212.

McRae., H. (1994). *The World in 2020*. HarperCollins, London.

McLoughlin., J. (1970). *Urban and Regional Planning*. Faber and Faber, London.

Meadows., D, D. Meadows and J. Randers (1972) *The Limits to Growth*, Universe Books, N.Y.

Miles., R. E. (1976). *Awakening from the American Dream*, Universe Books, N.Y.

Myers., N. (1990). *The GAIA Atlas of Future Worlds*. Robertson McCarta, London.

Naisbitt., J. (1994). *Global Paradox*. Allen & Unwin, St Leonards.

Naisbitt., J. (1996). *Megatrends Asia*. Nicholas Brealey Publishing Limited, London.

Newton., P. W. and M. A. P. Taylor (1985). Urban Futures: A Exploratory Study. CSIRO - Division of Building Research. Internal Report No. 85/2 R. (cited with permission of M. A. P. Taylor.)

Penrose., R. (1989). *The Empror's New Mind*. Vintage, N.Y.

Peters., T. (1994). *The Tom Peters Seminar*. Macmillan, London.

Popcorn., F. and L. Marigold (1996). *Clicking*. HarperCollins, N.Y.

Reichenbach., H. (1970). *Experience and Prediction*. Chicago, Univ. of Chicago Press, especially pp. 373-387.

Rheingold., H. (1992). *Virtual Reality*, Mandarin, London.

Rugge., J. G. (1975). 'Complexity, planning and public order', in T. R. La Porte (Ed.), *Organized Social Complexity - Challenge to Polities and Policy*, Princeton Univ. Press, Princeton.

Sagan., D. (1990). *Biospheres: Metamorphosis of Planet Earth*. Arkana, London.

Schaffer., J. D. (1985). 'Learning multiclass pattern discrimination', 1st International Conference on Genetic Algorithms, pp. 74-79.

Schaffer., J. D. (1987). 'An adaptive crossover distribution mechanism for genetic algorithms', 2nd International Conference on Genetic Algorithms, pp. 36-40.

Senge., P. M. (1990). *The Fifth Discipline: The Art & Practice of The Learning Organisation*. Random House. Sydney.

Shaefer., C. G. (1987). 'The ARGOT strategy: Adaptive representation genetic optimizer technique', 2nd International Conference on Genetic Algorithms, pp. 50-58.

Smith., S. F. (1984). 'Adaptive learning systems', In R. Forsyth (Ed), *Expert Systems, Principles and Case Studies*. Chapman and Hall, London.

Snow., C. P. (1961). 'What is the world's greatest need?' New York Times Magazine, 2 April.

Stadnyk., I. (1987). 'Schema recombination in a pattern recognition problem', 2nd International Conference on Genetic Algorithms, pp. 27-35.

Starkie., D. N. M. (1974). 'Transport Planning and the Policy-Modelling Interface', Transportation. Vol 3. No. 4. pp. 323-334.

Suh., J. Y. and D. V. Gucht (1987). 'Incorporating heuristic information into genetic search', 2nd International Conference on Genetic Algorithms, pp. 100-107.

Syswerda., G. (1989). 'Uniform Crossover in Genetic Algorithms', 3rd International Conference on Genetic Algorithms, pp. 2-9.

Tanese., R. (1987). 'Parallel genetic algorithm for a hypercube', 2nd International Conference on Genetic Algorithms, pp. 177-184.

Taylor., M. A. P., W. Young and P.W. Newton (1988). 'PC-based Sketch Planning Methods for Transport and Urban Applications', Transportation. Vol. 14. pg. 361-375.

Thiemann., H, (1973). Interview in Europhysics News, August.

Toffler., A. (1983). *The Third Wave*. Pan, London.

Toffler., A. (1991). *Power Shift*. Bantam, N.Y.

Tydeman., J. (1987). *Futures Methodologies Handbook: An Overview of Strategic research Methodologies and Techniques*. Commission for the Future, Renwick Pride Pty. Ltd.

Vickers., G. (1970). *Freedom in a Rocking Boat: Changing Values in an Unstable Society*, Allen Lane, Penguin, London. p. 43.
Wagar., W. (1971). *Building the City of Man: Outlines of a World Civilization*, Grossman, New York.
Waldrop., M. M. (1992). *Complexity*. Penguin, London.
Waterman., R. (1994). *Frontiers of Excellence*. Allen & Unwin, N.Y.
Wheatley., M. J. (1992). *Leadership and the New Science*. Berrett-Koehler, San Fransisco.
Wilson., S.W. (1985). 'Adaptive "cortical" pattern recognition', 1st International Conference on Genetic Algorithms, pp. 188-196.
Woodcock., A and M. Davis (1980). *Catasrophy Theory*. Penguin, London.

There is another author we must thank however we cannot put a name to him/her. The problem arises because the reference was collected a number of years ago at the Monash University Library while one of the authors was on a trip. Phototcopies of relevant parts of their works were taken, brought home, and included into the original draft of this book. It was then discovered that the photocopies had no information that made it possible to reference the authors. Rather than remove this information, which is of vital importance, it was decided to refer to the author in the manner you observe here. If the author of this information can come forward they will be recognised in any further issues. Table 6.1 is one of the items that need to be referenced in this manner as are the questions given on the previous page that relate to this table.